Science in Latin America

A History

EDITED BY JUAN JOSÉ SALDAÑA
TRANSLATED BY BERNABÉ MADRIGAL

University of Texas Press ⬥ *Austin*

Copyright © 2006 by the University of Texas Press
All rights reserved
Printed in the United States of America
First edition, 2006

Originally published as *Historia social de las ciencias en América Latina*
© UNAM and M. A. Porrúa, Librero-Editor, Mexico City, 1996

Requests for permission to reproduce material from this work should be sent to:
   Permissions
   University of Texas Press
   P.O. Box 7819
   Austin, TX 78713-7819
   www.utexas.edu/utpress/about/bpermission.html

♾The paper used in this book meets the minimum requirements of ANSI/NISO Z39.48-1992 (R1997) (Permanence of Paper).

Library of Congress Cataloging-in-Publication Data

Historia social de las ciencias en América Latina. English.
Science in Latin America : a history/edited by Juan José Saldaña ; translated by Bernabé Madrigal.— 1st ed.
   p. cm.
Includes bibliographical references
ISBN-13: 978-0-292-71271-3 (alk. paper)
ISBN-10: 0-292-71271-5 (alk. paper)
1. Science—Latin America—History.   2. Science—Social aspects—Latin America.
I. Saldaña, Juan José.   II. Title.
Q127.L38H57813 2006
509.8—dc22
2005035364

Science in Latin America

Contents

Introduction: The Latin American Scientific Theater   1
JUAN JOSÉ SALDAÑA

1. Natural History and Herbal Medicine in Sixteenth-century America   29
XAVIER LOZOYA

2. Science and Public Happiness during the Latin American Enlightenment   51
JUAN JOSÉ SALDAÑA

3. Modern Scientific Thought in Santa Fe, Quito, and Caracas, 1736–1803   93
LUIS CARLOS ARBOLEDA AND DIANA SOTO ARANGO

4. Scientific Traditions and Enlightenment Expeditions in Eighteenth-century Hispanic America   123
ANTONIO LAFUENTE AND LEONCIO LÓPEZ-OCÓN

5. Science and Freedom: Science and Technology as a Policy of the New American States   151
JUAN JOSÉ SALDAÑA

6. Scientific Medicine and Public Health in Nineteenth-century Latin America   163
EMILIO QUEVEDO AND FRANCISCO GUTIÉRREZ

7. Academic Science in Twentieth-century Latin America   197
HEBE M. C. VESSURI

8. Excellence in Twentieth-century Biomedical Science 231
   MARCOS CUETO

9. International Politics and the Development of the Exact
   Sciences in Latin America 241
   REGIS CABRAL

Science in Latin America

INTRODUCTION

The Latin American Scientific Theater

JUAN JOSÉ SALDAÑA

This volume collects for the first time a history of science as a whole in the geographical and cultural region known as Latin America. The authors are historians of science and discuss, among other issues, what, at different moments and under different circumstances, has been understood as science in Latin America, the forms scientific activity has taken, the settings responsible for the autochthonous peculiarities of science in the region, and the adoption of European science and its evolution in Latin America. This is a local history of how geographical accidents, individuals and groups of individuals, institutions, ideologies, concepts, and scientific theories affect one another in a specific social and cultural context.

This social history of science by no means scorns the intellectual aspects of science. On the contrary, it helps us understand the nature and behavior of social groups (the scientists) that created, developed, or incorporated concepts and theories in a particular social context and always as a consequence of it. Equal attention is paid to the general aspects of society and regional geography (the social order, culture, natural resources, geographical location, etc.) that are responsible for attitudes toward science and that have imposed a particular style on it. The authors in this volume use new analytical perspectives (forms of approaching the history of science) and offer a novel image of the Latin American scientific past.

Why has no such history been written before? Have we not had, for centuries, the most varied testimony that original experiments in science and technique were developed on this section of the planet? In addition, in practically the whole of Latin America, there have been significant efforts to record the history of moments, people, institutions, achievements, and other aspects of the scientific activity that has taken place there. There have even been histories that present science as formulating an entire national vision.[1] Nevertheless, it apparently has not been understood that science is one of many shared threads that will surely continue to tie together even more tightly the framework of Latin American history. Science, by being intertwined with other aspects of social and cultural life, has been the cause as well as the effect of important regional historical events and is viewed as such in the chapters that make up this book. Certainly, there are also national differences to be taken into account, dating back many centuries and developed in unequal circumstances.

The premise of the authors of the studies gathered here is that Latin America is a region. In the American geography of science, nontrivial aspects can be distinguished by understanding their nature and organization, since they are what makes this region different from others. Topographic relief, the nature of the terrain, the climate, the wide diversity of flora and fauna in some areas, the types of mineral deposits, and so on, create very defined and characteristic geographical areas, such as the Amazon region, the Mesoamerican lowlands, the Andean and Mexican highlands. The cultural and scientific patterns in the region depend on these material conditions.[2] Clear examples include Amerindian science and technology (herbalism, astronomy, agricultural techniques, medicine) and natural history from the seventeenth to the nineteenth centuries (botany, zoology, paleontology, mineralogy), as will be seen.

Likewise, characteristic collective features have sprung up throughout the historical dynamics of the region. Examples include highly developed autochthonous cultures and civilizations in the Andes and in Mesoamerica, which imposed, and still impose, particular features on those regions' social life. This circumstance determined, for instance, the European conquerors' decision to establish their main viceroyalties in Peru and Mexico—places where socially, politically, and culturally advanced societies already existed. In contrast, the Anglo-Saxon colonization of the northeastern portion of North America followed a totally different pattern. With an indigenous population that may be called scarcely developed from a cultural and social point of view, the social organization developed in that area imitated that of Europe and differed from that developed in indigenous-Hispanic-Portuguese America. Nevertheless, we should note that a particular style of scientific tradition also developed in the original thirteen British colonies through time and across social dynamics.[3]

Styles in science reveal the diverse sociohistorical and geographical conditions to which it has been subjected. Such is the case of Spanish America, where, as the Conquest and the imposition of Western civilization truncated native social and cultural processes, syncretism between old local knowledge and new knowledge imposed by Europeans, the cross-fertilization of those systems, and nontypical forms of European and Amerindian traditional scientific practice emerged. Proof lies in the survival of significant indigenous communities that possess a solid culture with elements of their traditional science and techniques.

The new societies that began to develop in the Americas in the sixteenth century led to historically unprecedented social and cultural evolution. New social characteristics derived both from the imposed social and cultural order and from the native one. For our purposes, the settlers' gradual creation of their own Latin American identity (by "Indianizing" the natives and through miscegenation and Creolization [*criollismo*]) during three hundred years of colonial life is especially significant. These social and cultural facts were expressed in the science that developed in the region. That science had a propensity to use established knowledge to understand the immediate natural and human reality or to develop it if necessary. A territorial, or telluric, feeling and a zeal for having their knowledge play a social role were important intellectual and social stimuli and acted on Latin American scientists at the end of the colonial era.

The emancipation of American societies in the second decade of the nineteenth century once again saw scientific activity take a particular course. The new American nations welcomed the nineteenth century optimistically, and their hopes lay with science to impart modern education to their people and to provide for the well-being and the material transformation of society. Colonial sociocultural inheritance and the formidable internal and external challenges that the new nations had to face impeded the scientific projects of the "Independence Generation." Nevertheless, Latin America felt the need to transform and adapt itself to the new requirements of its social evolution. Educational reform based on science was undertaken in all the newly independent countries, and the employment of science and technology for social and political ends, envisioned at independence and motivated by the goal of "progress," began in the second half of the century.

In the twentieth century, especially toward the end, science and technology became omnipresent. They embraced everything, including international affairs. Science also became an important academic discipline, and "invisible" schools communicate more intensively now. In the middle of the twentieth century, the need to encourage strategies for social and economic development that included science and technology as an important component became evident. Facing such critical situations as foreign indebtedness and the concomitant reduction of funds spent on science and technology, Latin Americans realized that cooperation and regional scientific collaboration could be the answer to the "lost decade" of the 1980s.

With globalization of all types and the challenge of the third Industrial Revolution, coordination of science and technology became an adequate and appropriate strategy for the future of Latin America. At present, there are numerous areas of science and technology in which regional and sub-regional cooperation exist. Issues vital to all countries—such as protection of the environment, communications and information science, energy, new materials, health, agriculture, biotechnology, scientific and technological development, basic sciences, and the teaching of science—are demanding closer cooperation among Latin American nations.[4]

The choice of endogenous science and technology in these countries, in the opinion of the contributors to this book, is historically irreversible. But at the same time, it is equally important to recognize that people belong to a particular sociocultural environment and that modernization is not transferable nor does it guarantee success. The decades-long prevalence of scientific and technical voluntarism confirms this. This is why sociohistorical studies of science and technology have become indispensable and allow us to understand the cultural dimension of science and technology, wherein the main change actors may submit science to their control (i.e., contribute a bit of the realism so needed by current development projects).[5]

**Historiographic Approaches:
The Emergence of "Science on the Periphery"**

The history of science as a discipline has developed conceptually and technically since the 1940s. It has moved from the level of a "literary genre" imposed in the eighteenth century (to relate the main episodes in the evolution of science) to scientific ideas and the external conditions that make science possible.[6] The absence of Latin America and, broadly speaking, of the periphery draws attention in specialized international literature on the technical evolution of the historiography of science. This does not mean that historical studies of the science of these countries or the outlying regions are absent. There have been some, and it is possible to talk about certain traditions—like those in Latin America. Rather, I refer here to the embarrassing situation that a Latin American historian of science faces: on the one hand, the assertion of the universal and positive character of scientific knowledge; on the other, the generally recognized contextual nature of scientific activity. This situation consists of "being" (as José Sala says when referring to Latin America) between the history of science and its philosophy.[7]

This important issue has come into focus since the 1980s.[8] The well-known controversy between Spanish and Latin American science,[9] which included racial overtones and separated cultures and regions from the ability to produce science, has been put aside.[10] Fact-based research has been required to demonstrate how fruitful and varied Latin American scientific activity has been, to show the existence of theoretical issues in the historiography of science essential to specific geographical and cultural regions, and, finally, to pull down the complex framework of ideas about what the history of science and science itself are and to locate in the philosophical field of historiography the serious and pressing questions of what has been something more than mere forgetfulness: the existence of Latin American science.

Furthermore, in Latin America, the history of science had been essentially a "secret" history, according to Elías Trabulse, or even "not told," as Marcos Cueto claims.[11] Indeed, this history remained hidden and underground, though it developed chronologically parallel to the political, social, economic, and cultural events that constitute the past of the Latin American people.

Now then, why has this "secret" history remained secret, and why have historians, as a rule, not studied it? It is because of the approach, the historiographic methods and theories that have prevailed until recently. Nevertheless, the evolution of the historiography of science permitted the field to open up to scientific activity previously excluded cultural regions, Latin America, in our case. Thus, individuals and circumstances, texts, institutions, practices, policies, and theories that never before were thought of—and with which this book deals—emerged in the work of the historians of science.

Historical studies about local science had, until the 1980s, a limited horizon and a distorted understanding of scientific activity. Basically, the works that strove to record Latin American scientific experience were laudatory histories, chronologies of events, and commemorative accounts, all of them revealing methodological vagueness and little understanding of the peculiarities of the field. Other works sought to develop a history of the "contributions" to universal science. An underlying Eurocentrism in historiography led efforts beginning in the nineteenth century to outline the history of Latin American science conceptually, which, surprisingly and ex hypothesi, excluded Latin America. Besides, Latin American "contributions" were indeed very scarce. This situation prompted contemporary historians of science to develop a perspective from which science would arise, as Shozo Motoyama says, "as a social process that could be understood even outside the European framework."[12]

All of this illustrates the marginal role assigned to Latin America, which, although it did not participate in the Scientific Revolution,[13] had in the last four and a half centuries contributed original scientific achievements to the framework supported by the great European advances. Our interest in the social history of Latin American science is due to this fact. It is a history connected to the culture and identity of the region, because the undeniably valuable science that developed there interacted with the social context and is inseparably linked to it. Moreover, this history is also concerned with the general history of science, since it relates the complex process of the spread of European science,[14] as well as its adoption in the receiving countries. This process, of course, is an important part of the science that originated in Europe, and to study it is to see in the mirror the reflection of the source. Most important are the causes and circumstances of this reflection (whether distorted or refined) of a beam that did not spread unaltered through geography or time.

If the reflected image is interesting for the general history of science itself, the plane of reflection, continuing the metaphor, is interesting for understanding the role science played in the receiving societies as well as the structural difficulties of establishing and consolidating it as national science. The history of science can show, thus, the dynamics of scientific culture and communities, the particular scientific ethos, schools of thought, the social mechanisms for valuing scientific work, the institutions, development policy, the setting up of teaching institutions. When that history is compared to the European pattern, the "perverse" effects and other social aspects of great importance, besides allowing us to understand Latin American scientific development, help us clarify our options.

Historiographic Mimesis

In the 1950s, Latin American science historians discovered local science as a product of local history. By that time, methodologies like economism and social analysis had appeared and sought to define the object of the history of science. Pioneering works like those by José López Sánchez (*Tomás Romay y el origen de la ciencia en Cuba* [Tomás Romay and the Origin of Science in Cuba], originally published in 1950), Fernando de Azevedo (*As ciências no*

Brasil [The Sciences in Brazil], 1955), and Eli Gortari (*La ciencia en la historia de México* [Science in the History of Mexico], originally published in 1963), opened a new window in the historiography of Latin American science. This resurgence of historiography, however, did not bring about the immediate reform of epistemological problems or analytical categories. In fact, an uncomfortable Eurocentrism, originating from a certain methodological mimesis, continued in the form of an apparent inability to define and to grasp the appropriate object of the history of Latin American science, in spite of incorporating social history in the analysis. A later generation of historians declared the universalist model of the "history of science in Latin America" to be insufficient. Neither the mere reference to social facts nor to history was enough to understand the nature of science in these countries.

The philosophical criticism of methodologies has shown the usefulness of an "epistemological watchfulness" in this discipline. Until the 1980s, this issue had not been approached by specialists, who considered it natural to follow their European colleagues when studying the scientific past of Latin America. It was, in fact, in methodological imitation that science and its essential conditioners looked, in different social contexts, for Latin American contributions to the scientific mainstream or the socioeconomic and cultural conditioners typical of modern European science. Although positivism and economism were encouraged for nationalist purposes (i.e., to find a place for Latin America in the history of science), the issue of specificity was ignored. As a consequence, a strange and paradoxical historical discourse was produced that aimed to understand the historical element of a geographically and culturally defined science according to universal outlines. The new history of science, fundamentally understood as social history, made it clear that not only the object of study but also the concepts for grasping it were specific.

Between the 1930s and the 1950s, expectations arose in the region regarding a quick development of science in Latin America as part of economic-development projects.[15] These projects conceived of science as a factor in development. Therefore, Latin Americans sought to graft onto or to inject scientific modernization into society and to create institutions and policies specifically dedicated to this purpose. The issue of the situation of science in Latin America and its possible future consequences therefore arose. Unlike previous historiographic methods, the new concern was centered on identifying the conditions that make possible—or that impede—scientific development in a certain sociohistorical context. Fernando de Azevedo, for example, was inspired by the Weberian sociology of culture. The work of

López Sánchez and Gortari, in turn, originated in the methodology provided by historical materialism.

Azevedo saw Brazilian scientific development as an integral part of Brazilian culture.[16] He wondered about the causes of his country's scientific backwardness and tried to identify the ailment and propose remedies. His work claims that the association of civil and religious power, characterized by the Counter-Reformation in Portugal, is the cause, because obscurantist cultural policies, necessary for economic exploitation, were imposed in the colony. It was not until 1837, during the regency, that the rupture with the Jesuit colonial tradition of teaching began. Later, there were sporadic scientific manifestations in the field of experimental research and applied science. Finally, in the twentieth century, the effects on urban growth of the development of trade and industry, mainly in São Paulo and Rio de Janeiro, led to the founding of the country's first universities (in the 1930s) and of other scientific centers—which had, until then, abandoned the dominant (literary and rhetorical) cultural model. Azevedo concluded that the ideas that determined the Brazilian cultural process originated in particular policies and economic and social forces; only the transformation of these could have a positive impact on the evolution of culture and, as a consequence, of science. The history of science would, according to Azevedo, help Brazilians become aware of the changes that the national culture required for scientific development.

This analytical approach constituted a historiographic novelty when it appeared. For the first time, Brazilian scientific activity was related to significant social events in the country's history. But did it really give an account of scientific activity in the country? Azevedo's argument seemed to be a plea for what should be changed so that science could flourish. What he suggested was a different cultural framework capable of engendering science as it functioned in scientifically advanced countries.

Paradoxically, the thesis that appropriate conditions have not existed (i.e., the absence of the Protestant ethic and culture) for nurturing science contradicts the efforts of Azevedo and his collaborators to create a history of something that did not yet exist: science in Brazil. The illogicality can be explained by not excluding contextual treatment of the object of study and by considering it, rather, as an example of the prototype—European science. Appearances to the contrary, Azevedo's approach was a variation of Eurocentrism and, from the epistemological point of view, of externalism.

Inspired by Marxist theses, José López Sánchez (in Cuba) and Eli Gortari (in Mexico) composed historical interpretations of the scientific development of their countries, although López studied only the late eighteenth century.[17] Both authors maintained that the introduction of modern science was the result of establishing capitalism. In their opinion, the revolutionary effects of science were manifested in the economy because of science's direct contribution to the development of productive forces—on the ideological plane, by science's fight against religion and Scholastic thought; in education, by the progress science represented for agriculture and industry.

As several researchers have pointed out,[18] the supposedly productive function of science did not take place during the time these authors are talking about, in Cuba or in Mexico. There are also doubts about their anti-Scholastic and antireligious thesis, because it is not difficult to prove the union between science and faith in colonial Latin America. Regarding the educational impact of science, we must recognize that in Mexico, in contrast to other places, it was present but scarcely influenced industrial activities.

If we accept the theoretical model adopted by López Sánchez and Gortari, we would have to conclude that in Latin America there was no science, since the conditions considered necessary by proponents of externalism for the emergence of modern science were absent. Historians of European science who regularly avoided speaking of Latin American science came to this conclusion, in fact. López Sánchez and Gortari were inspired by the commendable intention of finding a historical place for Latin American scientific activity. It is, in fact, impossible if we accept economism and Eurocentrism. Accordingly, one has either to deny the existence of science in the periphery or qualify it as "exceptional" for lacking a theoretical locus.[19] Neither option gives a faithful account of what really happened, and from the historical point of view, neither answers the initial question: What has Latin American scientific activity been specifically, and what social conditions made it possible?

Such is the case with Azevedo's sociologism and López and Gortari's economist externalism, which involves a simple transfer of European historiography, which is positivist, reductionist, and ahistorical when applied uncritically to Latin America. It is, as a consequence, an inadequate methodology because it arises from mimesis. When the essential contextual differences between Europe and Latin America are not kept in mind, it is presumed that the same social factors act on the dynamics of science. To imitate in history is, therefore, to lose one's identity.

"Thinking Our Science": Twenty-five Years of the Historiography of Latin American Science

Since the mid-1980s, we have understood that the history of science and technology pose problems that are epistemological in nature (their geographical and cultural specificity, for example) and that the historian should not ignore. In fact, the discipline began to become self-aware when it began to ask what the history of science was about if it included contextual definitions? In this section, we will talk about how this self-awareness emerged.

To break with the methodological mimesis, it was necessary to chart a new course in historiography, one whose focus was the specificity of science in the periphery. "Thinking our science" became the watchword. This process began to take place in the 1980s and is characterized by conceptual and terminological modernization. It derives, however, from the progress of this discipline in other parts of the world—while Latin American historiographers move conceptual problems to the forefront as part of their search for an alternative to the internalism-externalism dilemma. Thus, the criticism of the perspective that focused the history of science on the history of scientific ideas and mathematics-based ideas is important, as it excluded the regions that were not in the so-called mainstream of scientific development. In the historiographic tradition, areas and cultures like those of Latin America and even the United States until the end of the nineteenth century were outside the field.[20] It therefore became necessary to break the schematism characteristic of externalism.

The founding of the Sociedad Latinoamericana de Historia de las Ciencias y la Tecnología (Latin American Society for the History of the Sciences and Technology) in the City of Puebla, Mexico, in 1982, and the launch of its journal, *Quipu, Revista Latinoamericana de Historia de las Ciencias y la Tecnología* in 1984, are equally important. Both helped spur growth in the 1980s of amateur approaches to making the discipline professional and to bringing international recognition to the history of science in the region.

Thus was developed a new language able to name situations and scientific actions never before considered by historians of science. It is therefore appropriate to say that, from that point, the scope of historians of science expanded. "Peripheral" science offered new, specific facets for historical study. In fact, the science of Latin America came to be recognized as science in its own context. Let us consider some of the main developments that have taken place in this discipline.

Long stages of Latin America's history, for well-known reasons, have been framed by religious culture. How did scientific thought and religious culture cohabit in Hispanic-Lusitanian America? From the traditional point of view, there was absolute opposition between science and religion. Therefore, for colonial scientists, science did not quite fit into this kind of interpretative scheme, and historians often denied its existence or, when it could not be ignored, declared scientific and technical discoveries from the Americas exceptional. But when the ties with traditional history were broken as a consequence of continued research, unsuspected individuals, scientific texts, and institutions emerged, such as López Piñero has made evident in the case of Spanish science.[21] Trabulse studied the exact sciences in sixteenth- and seventeenth-century Mexico, periods declared by previous historians to be void of science.[22] Along the same lines, Trabulse, in *Ciencia y religión en el siglo XVII*, studies astronomy and religion and declares that they were "indivisible, since focusing on only one aspect . . . would mutilate what should be considered as a whole."[23] In fact, this move was something more than uniting disparate terms, because studying comets in the sixteenth and seventeenth centuries necessarily meant mixing two conceptions of the universe: one coming from medieval science, and the other from the new mechanistic cosmology, which had come into existence by that time. It is worth pointing out that this happened not only in New Spain but also in Europe, so we can say that Latin American scientists were up on the scientific currents and concerns of the time.

The sociohistorical analysis of science was developed partly under the influence of functional sociology. Simão Schwartzman applies this methodology to the study of the formation of the Brazilian scientific community. Contrary to the idea that establishes science as a progressive construction of the ideal building of truth, and relying on R. K. Merton and Thomas S. Kuhn, Schwartzman studies important subjects such as science on the periphery and science and development. He establishes that the history of science can be seen "as the history of efforts to establish national scientific communities to work with the models, themes, and styles of work characteristic of science in each era."[24]

The work of Thomas S. Kuhn should be mentioned here because several scholars apply Kuhnian methodology to Latin American science. Trabulse does this in *Historia de la ciencia en México* when he assigns a pivotal role to the notions of "paradigm" and scientific community. This concept leads him directly to question the theoretical locus of Mexican science. By recognizing that scientific communities are characterized by their adherence to a paradigm, Trabulse is able to assign a locus to Mexican scientific activity, since, in his opinion, the latter exhibits successive adoptions of scientific paradigms, each one corresponding to the evolution of science. Trabulse also uses a complementary thesis about the continuity of Mexican scientific communities as a basis for asserting the existence of a local scientific tradition beginning in the sixteenth century.[25] He divides the dynamics of scientific development into periods corresponding to the changes in paradigms.

The notion of a scientific community is useful when applied to "normal science," such as the systems and institutions in which science is taught and those in which research is conducted. J. M. Carvalho's work on Brazil, Hebe Vessuri's studies on Venezuela's scientific institutions, J. I. López Soria's history of Peru's Universidad Nacional de Ingeniería (National Engineering University), María A. M. Dantes's works on research institutes in Brazil, and Nancy Stepan's study of the Instituto Oswaldo Cruz (among others) use this approach.[26] All of these authors are interested in understanding the mechanisms that have worked to institutionalize science in Latin America. This issue goes to the heart of the development of Latin American science because, according to Roger Hahn, scientific institutions are the forge in which knowledge and politics are joined to create viable science.[27]

Politics as a form of reconciling interests has become a research area in recent years in the more general sense of the role played by the state with regard to science, since the state acts on institutional mechanisms, organizational forms, goals, financing, and conditions that make scientific activity possible. After noticing that the fundamental factors for European scientific development (i.e., industry and the army) had virtually no influence on Latin America's scientific development, some researchers have wondered about Latin American science's agent of structural change. The very makeup of Latin America's history, that is, a strong political regime during the colonial period and an environment of economic exclusion since independence, has left it up to the nation-states to create a scientific infrastructure. The relationship between science and political order is by no means secondary, because it has had a direct effect on the conditions that made Latin American scientific development possible. This was Frank Safford's method for analyzing the attempts to form a technological elite in Colombia in the nineteenth century. Since there was no real industrial demand, this elite ended up in public administration or teaching, which thwarted technological advancement. Marcel Roche, by studying Rafael Rangel and the unavoidable problems he faced, has analyzed how Venezuelan science began to be politicized.[28] In Mexico, there have been studies that illustrate the coordinating role of the state regarding science since independence. I have shown how the policies of the state have been pivotal in the organization and promotion of scientific activity and, reciprocally, how science and technology have aided the nation-state in transforming politics into social engineering to create a new society. To the political figure of the "sovereign state" I have added the scientific figure of the "rational or scientific state."[29]

Latin American scientists have certainly maintained contact with Europe since the sixteenth century. This fact has led various authors to focus on science's transcultural transmission. This approach means analyzing processes of incorporation and the adoption of science in defined sociohistorical contexts. Lewis Pyenson, for example, has introduced new analytical categories such as "functionaries" and scientific "seekers." These have allowed us to clarify the complex structure, not foreign to international politics, in which scientific activity in Latin America has remained entangled. Such was the case of the exact sciences, subjected to the cultural imperialism from which Latin America has frequently suffered. Moreno and Pruna have studied Darwinism within this framework.[30]

The issue of the spread of science has produced very interesting results regarding our understanding of ways for adopting or appropriating theories and forms of scientific practice in different contexts. It is not possible to separate scientific diffusion on the periphery from the reception process, which is not only material but also conceptual, ideological, and cultural. In this sense, the diffusion-reception process was conceived as a historical reality rooted in social interests. Luis Carlos Arboleda has used concepts taken from the sociology of knowledge for analyzing the spread of scientific theories and established the following: that historical facts have been in charge of socializing scientific paradigms, that is, would be responsible for understanding the social and cultural conditions under which the original theoretical system underwent an intermediation and reinterpretation process until public opinion was favorable. José Sala and Antonio Lafuente have pointed out the importance of the socioprofessional role played by scientists in colonized regions for comprehending the institutionalization of science in New Spain. I, in turn, have shown that all diffusionist approaches to the history of science deliberately omit local context as an explanatory category for what happened in particular sociohistorical circumstances; thus, they make the science of peripheral areas dependent on the mythical and, consequently, ideological scientific misoneism of a "disinterested" Europe.[31]

The cultural role science has played in underdeveloped countries is also relevant to transcultural diffusion. If, as the historiography of European science has shown, there is a direct link between science and European civilization, then it is necessary to accept science as culture. This leads to an analysis that does not differentiate between life and its problems or between life and the intellectual objective of improving knowledge. Vessuri has pointed out that Latin American scientific knowledge can no longer be separated from the network of cultural constrictions or ideological commitments that ordinarily shape social and political decisions. The revaluation of local common sense and the (re)construction of traditions, therefore, become a very important objective for the researcher. In this same vein, Marcos Cueto has carried out research on the relationship between scientific research and nationalist and indigenist ideologies, as in the case of biomedical research in Peru.[32]

The question of the spread of science from a center identified with Europe during most of Latin America's history has begun to demand a different type of analysis, that is, multilateral analysis, since this is a truly complex matter. The belief that "the cause of everything" in Latin American scientific and technological processes is the diffusion of new theories and scientific methods and the organization of scientific activity from the home countries. This point of view assumes that cultural renovation is the expression of a single, fixed, and complete will working from the top down that diminishes until it finally disappears into the characteristic dynamics of "outlying" or colonial societies. For that reason, the topic of nationalism in science has begun to draw attention. I have shown that it is not possible to conceive of a scientific evolution of peripheral regions only as the result of an "injection" or "abrupt and quick introduction of science and technology."[33] We must take into account local factors, which usually constitute the sine qua non for globalizing science (its "ecology").

As we have seen, conceptualizing peripheral science as exceptional does not help us understand its process. Arboleda, following this reasoning, has called "patriotic science" those projects that were intended to help New Granada's elite prosper by using modern science for exploiting the country for the gain of local interests.[34] The concept of "national science" has expanded because it not only includes the national period but, in some cases (New Spain, Peru, and New Granada), it also takes in the end of the colonial period as well, when local science achieved a clearly central role in society (see Chaps. 2, 3, and 5 in this volume). It is worth emphasizing that colonial Spanish scientific and institutional initiatives relied on the national science that was flourishing in Mexico City, Lima, and Santa Fe de Bogotá.[35]

In the case of colonial Latin America, and particularly in regions like Mexico and Peru, the notion of "scientific tradition" or "endogenous science" (sometimes, of course, not free from social exclusion or heterodoxy) must be introduced in order to understand the cultural, ideological, and political context in which the spread of science—which generates social and cultural effects in the receiving country—takes place. Indeed, the failure of projects conceived in the home countries, polemical discussions with foreign scientists, and the perverse effects of foreign science occurred more frequently than is generally imagined. A quick look at the causes is enough to demonstrate that the spread of science does not take place in a cultural vacuum; for that reason, there may be opposition to it or to development modalities—such as happened in the Mexican case with chemistry, physics, and botany, or in Colombia in relation to mathematics.[36] For this reason, it has been necessary to look for the roots and the context that made imported and local science in peripheral regions historically possible.

In the case of New Spain, the existence of ideological nationalism and a "Creole" scientific tradition shows, for example, that in this colony there was an appropriate foundation for introducing scientific theories, institutions, and policies at the end of the eighteenth century. That is, Mexican science followed an evolutionary organizational process that ended up, naturally, in institutionalization.[37] In other words, there was an accumulative process of social experience regarding scientific knowledge and its organization. Until now, excessive weight has been given to external elements and their concomitant ruptures, because the relationship between external and internal components in Latin American science has not been treated as a dialectic.

Finally, I must point out that in this analysis it is increasingly crucial to study the role played in Europe by scientists born in the periphery and the influence they exerted on science in the home countries.[38] This will help us better understand the "bidirectional" action of the process of spreading science.

The increasingly clear placement of Latin American science within a defined social context does not mean that interest in "scientific matters" or in the regional peculiarities of knowledge production has disappeared altogether. The emergence of conceptual tools for including science in the Latin American context was so innovative that it was necessary to use borrowed terminology before it was possible to produce an adequate conceptualization and expression. The notion of periphery applied to science was born in this way.[39] The "periphery" category originated in dependency theory (prevalent during the 1960s and the 1970s) and dealt with the structural and asymmetrical relationship between the industrialized and the underdeveloped countries. When applied to science, the concept of asymmetry was used to compare underdeveloped countries with those considered as knowledge-producing centers and owners of scientific infrastructure, communities, and a research tradition. "Peripheral science" turned out to be the science produced in countries with a small structured scientific community, where just a tiny part of their gross domestic product was directed to scientific development and scientists' productivity was low (measured in terms of how many times their articles were quoted in international specialized journals). At present, the validity of several conclusions derived from the notion of peripheral science is being questioned.

Other conclusions express realities that are impossible to ignore because they refer to fundamental conditions such as the small investment in science and technology and the small number of researchers per capita. In particular, the criteria of productivity and other so-called international indicators of quality in scientific activity have been challenged because of their biased nature and because they do not correspond to the regional modalities of scientific practice.[40]

Historians of science have pointed out how rigid the concept of dependence is when it is applied to historical events, since it prevents the understanding of facts and situations. How do we explain, for example, that sometimes Latin American science has been "central" in relation to European science? New Granadan botany and Mexican herbalism, colonial Peruvian mathematics, New Spanish metallurgy, or the fact that Newton and Sigüenza were contemporaries—all of these are examples of past scientific excellence. More recently, we have discoveries in endocrinology by Argentine Bernardo Houssay and in microbiology by the Brazilian Oswaldo Cruz. In the 1980s, Marcos Cueto studied "the strange combination of modern and creative work in supposedly traditional and 'peripheral' contexts that are far from the world centers of science" and showed that biomedical research in Peru in the first half of the twentieth century displayed academic excellence.[41] Roy Macleod, on the other hand, has introduced the idea of the "moving metropolis" to refer to the intellectual creativity that takes place in colonial regimes inside a dialectical alternation between diffusion and reelaboration: this concept is today drawing the attention of researchers worldwide because, in some measure, it also characterizes European science.[42]

It is precisely the lack of original thought, since there is always a debt to the pioneers in a discipline, that raises the following question: When is a scientific community really creative? In a sense, all communities are receptors but not all communities reprocess what they receive until they have made it their own and contributed to its development. "Science is dependent when it passively receives and does not reprocess," as Celina Lértora writes. This Argentine historian has defined reelaboration as a characteristic of Latin American scientific activity in the following terms: there is reelaboration when the research in the receiving community produces a theoretically (or technically) different result from the previously held assumption, and this result is obtained independently of other scientific communities.[43]

Regarding the conditions that render possible the creativity and endogenous development of science, Shozo Motoyama describes Latin American scientific evolution as a process that advances by phases using a model called MTS (mental and technical substratum).[44] The first phase is the mental substratum, or "intersubjective base that provides and limits the action of intellect to formulate questions about nature." The second phase, the technical substratum, is "able to engender the instruments and necessary apparatuses to capture the answers to the questions already posed about nature—arising out of the mediation of the mental substratum."[45] The maturation of both phases permits full scientific development. Motoyama analyzes the case history of physics in Brazil and studies the formation of mental and technical substrata. He concludes that the dynamics of scientific, technological, and cultural development in his country since the 1970s has made these substrata reach a certain level of development, which has rendered possible for Brazilian science the end of a merely imitative stage and the beginning of another, more authentic and original, one.[46]

Although this review of the conceptual progress made in the discipline called the social history of science in Latin America is superficial, it nevertheless allows us to appreciate the remarkable dynamism of this branch of history in recent decades and how it has rightly become a field of knowledge. In addition to the abovementioned studies, there are abundant empirical ones that represent an interesting rereading of the scientific past and, not infrequently, offer traditional historiography new or formerly ignored data. An image, to a certain extent unexpected, of Latin American science and its social and cultural relationships has emerged but remains, in great measure, unknown outside the scholarly environment.

This book was organized to remedy, at least in part, this situation. A select group of researchers was invited to introduce their work to the public. These studies have been carried out using new theoretical approaches to the history of science in context. Our intention is to analyze, for the first time, both comparatively and comprehensively (chronologically and thematically), the scientific experience of this region, so vast in both time and space. The result is a group of monographs aimed at the learned public interested in discovering the dimensions and the value of Latin American science.

The book also contains several innovations. Concerning the colonial period, Chapters 2, 3, 4, and 5 introduce novel ways of referring to the scientific activity that took place during the Enlightenment. In Chapter 2, I consider scientific modernization to be the product of the social dynamics (economic, demographic, cultural, technical, and scientific) that the region underwent and also an answer to the needs brought about by such development. By abandoning traditional explanations wherein modernization relies on external factors (such as "good king" Charles III, the progressive ideas of the transnational Society of Jesus, and the misoneism of "scientific centers"), I clearly demonstrate how nationalism was imposed on Enlightenment science and its liberating ideological character.

Luis Carlos Arboleda and Diana Soto Arango (Chap. 3) analyze eighteenth-century polemics regarding the introduction of modern scientific theories and discover that these arguments were motivated by the struggle between the civil and the religious sectors for cultural and educational control, and not by "disinterested" actors in the modernization process—as has sometimes been claimed. They demonstrate the inadequacy of the customary diffussionist explanation, which omits the local context in which modernity is received and adopted and which has also always left its imprint on the final result.

Antonio Lafuente and Leoncio López-Ocón (Chap. 4), likewise, leave behind the view that presents initiatives for scientific expeditions as arising exclusively in the home countries. These authors also recognize strong local roots—viceregal and ecclesiastical—as two of the main sources for information about nature, geography, and American communities. They also highlight the process of cultural regionalization that originated in expeditionary scientific activities, particularly those in the eighteenth century.

The birth of the independent American nations stimulated unforeseen scientific activity. Contrary to the traditional thesis that independence created a scientific void, in Chapter 5, I refer to the emergence of nation-states simultaneously with an awareness of the value that science would have for completing political emancipation. In an atmosphere of prevailing optimism about what newly won freedom would involve, the independence-era generation formulated in all countries a republican scientific ideology, which received social and political sanction by being included in the constitutions of the new states. The nineteenth century began with this ideology on the horizon, and throughout the century diligent efforts were made to turn it into a reality.

Emilio Quevedo and Francisco Gutiérrez (Chap. 6) study medicine and public health in the nineteenth century. As happened in other areas, during the century, medicine underwent deep conceptual and technical transformations. With frequent references to, for example, Bogotá, Rio de Janeiro, Lima, Mexico City, and Buenos Aires, the authors describe the processes by which medicine was "modernized" and point out the role local context played in shaping theoretical and practical medicine capable of responding to public-health needs.

Hebe M. C. Vessuri (Chap. 7) presents the emergence of academic science in the twentieth century and the internal and external factors that contributed to it. It is worth mentioning that historians devoted little attention to twentieth-century science before Vessuri. Vessuri uses a sociological thesis and periodization for the purpose of understanding the nature and the complexity of the evolution of academic scientific activity in Latin America during most of the twentieth century.

Marcos Cueto (Chap. 8), also referring to the twentieth century, presents the case of the successful maturation of a scientific discipline in Latin America, namely, physiology, and the role new historiography can play in dissolving and abandoning prejudices (as ingrained as they are frequently false) that exist equally among the region's scientists and its historians. False conceptions include, among others, the dependent character of Latin American science, the necessity of big investments for its development, and the invariably negative influence of nationalism on science.

As is well known, the First World War involved scientists in a very significant fashion, and from then on, their activities acquired political and international relevance. Regis Cabral (Chap. 9) studies, by looking at major events, the less well known but significant case of the exact sciences in Latin America and their involvement in worldwide political problems. If we want to understand completely the diverse endogenous and exogenous factors that are involved in the field of regional science, we must make Latin American and world diplomatic history another element in the social history of science.

I believe this book presents new perspectives on Latin American science. Knowledge and the wide diffusion of the region's scientific history becomes indispensable when, as now, science and technology have reached a significant level of development. It is now when scientific efforts must be increased considerably in order to respond to the enormous challenges resulting from the region's underdevelopment. This step is necessary for forming a regional or local scientific culture and for continuing to build the "ecological niche" that the science of Latin America needs.

It is evident that very important aspects, such as strategic viability, need pertinent historical information to allow the main agents of change to modify and control themselves. It is equally important to base these modifications on the social history of local science. In short, it is fundamental "to learn from oneself." Thus, for that purpose, the history of science must be written to allow the construction of national scientific capability. In Latin America, the collective amnesia about the scientific and technological past cannot continue without the risk of losing the future.

As the recent European and Asian experience has proved, the cultural diversity of those regions that make up the global "marketplace" has become (in fact, it always was) the most valuable resource. Such diversity constitutes the reserve of experiences from which we will have to start creating true international science and society. To know and comprehend it is, therefore, imperative.

Postscript, 2001

The Spanish-language version of this book was published at the end of the twentieth century. This translation (financed by the Universidad Nacional Autónoma de México [National Autonomous University of Mexico]) will appear at the dawn of the new century, after the XXI Congreso Internacional de Historia de la Ciencia (International Congress of the History of Science) in Mexico City. That conference's general theme was "Science and Cultural Diversity." This English-language version allows the conference's findings to find worldwide distribution. In this way, we hope to create a new image of science as all-inclusive, in order to see the universal, as Portuguese novelist Miguel Torga says, as the local without walls.

Notes

1. See Babini, *La evolución*; Condarco, *Historia del saber*; Guimarães and Motoyama (eds.), *História das ciências*; Gortari, *La ciencia*; Yepes (ed.), *Estudios*; and several studies published since 1984 in *Quipu, Revista Latinoamericana de Historia de las Ciencias y la Tecnología*.
2. On the importance of geographical determination to science in all times and places, see Dorn, *The Geography of Science*.
3. See Reingold, *Science, American Style*. Regarding the birth of national styles in other regions, see Reingold and Rothenberg (eds.), *Scientific Colonialism*; Numbers (ed.), *Medicine*; Petitjean, Jami, and Moulin (eds.), *Science and Empires*.
4. See Saldaña (ed.), *La ciencia y la tecnología*.
5. See my contribution to the symposium "Science et développement?" in Petitjean and Jami (eds.), *Science and Empires*, pp. 380–383; Salomon, Nakayama, et al. in ibid.
6. See Russo, *Nature*; Saldaña (ed.), "Fases principales," pp. 21–78.
7. Sala Catalá, "La ciencia iberoamericana." Sala goes on to point out the following: "Neither formerly nor at present have observation, theories, or experiments been planned using any logic concomitant with the truth of language—on the contrary, their relations, both casual and ordered interaction, become meaningful in reference to a specific geography and history" (all translations are mine unless otherwise noted).
8. See "Actas"; "Historia y filosofía."
9. See García Camarero and García Camarero, *La polémica*.
10. Regarding European contempt for American nature, people, and knowledge, see Gerbi, *La disputa*.
11. Trabulse, "Introduction"; Cueto, *Excelencia científica*.
12. Motoyama, "História da ciência," p. 172.
13. See López Piñero, *Ciencia*; see also Peset (ed.), *La ciencia moderna*.
14. Regarding transcultural transmission of science, see Saldaña (ed.), *Cross Cultural Transmission*; Pyenson, "Functionaries"; idem, "*In Partibus Infidelium*."
15. Sagasti, "Esbozo histórico," pp. 31ff.
16. Azevedo (ed.), *As ciências no Brasil*; see esp. the introduction.
17. López Sánchez, *Tomás Romay*; Gortari, *La ciencia*.
18. See Moreno Fraginals, *El ingenio*, vol. 1, p. 132; Saldaña, "Marcos conceptuales"; idem, "The Failed Search," pp. 33–57.
19. As Gortari describes the rich and varied scientific works in seventeenth-century Mexico, or what he calls mere "anticipations" in reference to technical inventions. See Gortari, *La ciencia*, pp. 197 and 225, and my critique in "Marcos conceptuales," pp. 76–78.

20. See Reingold (ed.) *Science*, p. ix.
21. See López Piñero, *Ciencia*.
22. One of the first formulations of this thesis was presented by positivist historians of nineteenth-century Mexican science. See *Historia de la medicina*; Parra, "La ciencia en México," vol. 2, pp. 1900–1902.
23. Trabulse, *Ciencia*.
24. Schwartzman, *Formação*, p. 24.
25. I critique Trabulse's use of the notion of "scientific community" in "Historia"; see also Lafuente, "La ciencia desvelada."
26. Carvalho, *A Escola de Minas*; Vessuri (comp.), *Las instituciones científicas*; López Soria, *Historia*; Dantes, "Institutos," pp. 341–380; Stepan, *The Beginnings*.
27. Hahn, *The Anatomy*.
28. Safford, *The Ideal*; Roche, *Rafael Rangel*.
29. Saldaña, "La ciencia y el Leviatán mexicano"; see also idem, "Acerca de la historia."
30. Pyenson, *In Partibus Infidelium*; idem, "Functionaries"; the debate published in *Quipu*, vols. 2 and 3 (1985 and 1986), among Stepan, Safford, Glick, and me; Moreno, "México"; Glick (ed.), *Comparative Reception of Darwinism*; Pruna, "La recepción."
31. Arboleda, "La difusión científica"; idem, "Sobre una traducción"; Lafuente and Sala, "Ciencia colonial"; Saldaña, "Nacionalismo."
32. Vessuri, "Los papeles culturales," pp. 7–18; Cueto, "Nacionalismo."
33. Saldaña, "Nacionalismo," p. 120; idem, "Acerca de la historia." This position has been sustained by diffusionist theory. See, for instance, Peset, *Ciencia y libertad*.
34. Arboleda, "La ciencia," pp. 193–225.
35. Saldaña, "Acerca de la historia."
36. Aceves, "La difusión" (Mexico); Ramos, "La difusión" (New Spain); Arboleda, *Matemáticas* (Colombia).
37. See Saldaña, "The Failed Search," and Chap. 2 here.
38. For instance, José Mariano Mociño of New Spain, who directed the Jardín Botánico (Botanical Garden) in Madrid, or José María de Lanz, also from New Spain, who obtained in Paris a high position in the sciences and who wrote, for the Polytechnic School (with A. De Betancourt), a fundamental technology text: *Essai sur la composition des machines* (1808). On Mociño, see Lozoya, *Plantas*; on Lanz, see García-Diego, *En busca*.
39. See Sagasti, "Underdevelopment"; Sábato (ed.), *El pensamiento latinoamericano*.
40. For a critique of the limitations of bibliometric indicators such as the *Science Citation Index*, see Arvanitis and Chatelin, *Strategies scientifiques*.
41. Cueto, *Excelencia científica*, p. 29.
42. Macleod, "De visita," pp. 217–240.
43. Lértora Mendoza, "Un problema metodológico," p. 157.
44. Motoyama, "Algumas reflexões."
45. Motoyama, "Un análisis," p. 49.
46. Motoyama, "A física," pp. 61–92.

Bibliography

Aceves, P. "La difusión de la química moderna en México en el Real Jardín Botánico de la Ciudad de México." MA thesis, Universidad Nacional Autónoma de México, 1989.

"Actas del Seminario Internacional sobre Metodología para la Historia Social de las Ciencias en América Latina." In *Ciencia, técnica y estado en la España ilustrada,* J. Fernández and I. González (eds.). Madrid: Ministerio de Educación y Ciencia, 1990.

Arboleda, L. C. "La ciencia y el ideal de ascenso social de los criollos en el Virreinato de Nueva Granada." In *Ciencia, técnica y estado en la España ilustrada,* J. Fernández and I. González (eds.). Madrid: Ministerio de Educación y Ciencia, 1990.

———. "La difusión científica en la periferia: Newton." *Quipu* 4, no. 1 (1987): 7–32.

———. *Matemáticas, cultura y sociedad en Colombia.* Bogotá: Colciencias, 1986.

———. "Mutis entre las matemáticas y la historia natural." In *Historia social de las ciencias: Sabios, médicos y boticarios,* D. Obregón (ed.). Bogotá: Universidad Nacional de Colombia, 1987.

———. "Sobre una traducción inédita de los *Principia* al castellano hecha por Mutis en la Nueva Granada *circa* 1770." *Quipu, Revista Latinoamericana de Historia de las Ciencias y la Tecnología* 4, no. 2 (1987): 291–313.

Arvanitis, R., and Y. Chatelin. *Strategies scientifiques et développement: Sol et agriculture des regiones chaudes.* Paris: Editions de l'ORSTOM, 1988.

Azevedo, F. de (ed.). *As ciências no Brasil.* 2 vols. São Paulo: Edições Melhoramentos, 1955.

Babini, J. *La evolución del pensamiento científico en la Argentina.* Buenos Aires: La Fragua, 1954.

Carvalho, J. M. *A Escola de Minas de Ouro Petro: O peso da glória.* Rio de Janeiro: Editora Nacional, 1978.

Condarco, R. *Historia del saber y la ciencia en Bolivia.* La Paz: Academia Nacional de Ciencias de Bolivia, 1978.

Cueto, M. *Excelencia científica en la periferia: Actividades científicas e investigación biomédica en el Perú, 1890–1950.* Lima: Grade-CONCYTEC, 1989.

———. "Nacionalismo e investigación biomédica en el Perú." *Quipu* 4, no. 3 (1987): 327–356.

Dantes, M. A. M. "Institutos de pesquisa científica no Brasil." In *História das ciências no Brasil,* M. Guimarães and S. Motoyama (eds.). São Paulo: EDUSP, 1979.

Díaz, E., et al. *La ciencia periférica: Ciencia y sociedad en Venezuela.* Caracas: Monte Ávila, 1983.

Dorn, H. *The Geography of Science.* Baltimore, Md.: The Johns Hopkins University Press, 1991.

Flores, Francisco. *Historia de la medicina en México, desde la época de los indios hasta el presente.* 3 vols. Mexico City, 1886–1888.

García Camarero, Ernesto, and Enrique García Camarero. *La polémica de la ciencia española.* Madrid: Alianza, 1970.

García-Diego, J. A. *En busca de Betancourt y Lanz.* Madrid: Castalia, 1985.
Gerbi, A. *La disputa del Nuevo Mundo.* Mexico City: FCE, 1960.
Gortari, E. *La ciencia en la historia de México.* Mexico City: FCE, 1963.
Guimarães, M., and S. Motoyama (eds.). *História das ciências no Brasil.* 3 vols. São Paulo: EDUSP, 1979.
Hahn, R. *The Anatomy of a Scientific Institution: The Paris Academy of Science.* Berkeley & Los Angeles: University of California Press, 1969.
"Historia y filosofía de las ciencias en América." XI Congreso Interamericano de Filosofía, Guadalajara, Mexico, 1985. In *El perfil de la ciencia en América,* J. J. Saldaña (ed.). Cuadernos de Quipu 2. Mexico City: Sociedad Latinoamericana de Historia de las Ciencias y la Tecnología, 1987.
Lafuente, A. "La ciencia desvelada: Marginalidad y reinvindicación de la ciencia en México." *Asclepio* 37 (1985): 399–405.
———, and J. Sala. "Ciencia colonial y roles profesionales en la América española del siglo XVIII." *Quipu* 6, no. 3 (1989): 387–403.
Lértora Mendoza, C. A. "Un problema metodológico de la historia de la ciencia latinoamericana: Recepción vs creación." In *Historia social de la ciencia: Sabios, médicos y boticarios en Colombia,* D. Obregón (ed.). Bogotá: Universidad Nacional de Colombia, 1987.
López Piñero, J. M. *Ciencia y técnica en la sociedad española de los siglos XVI y XVII.* Barcelona: Labor, 1979.
López Sánchez, J. *Tomás Romay et l'origine de la science à Cuba.* Havana: Institut du Livre, 1967.
López Soria, J. I. *Historia de la Universidad Nacional de Ingeniería, 1876–1909.* Lima: Universidad Nacional de Ingeniería, 1981.
Lozoya, X. *Plantas y luces in México: La real expedición científica a Nueva España (1787–1803).* Barcelona: Ediciones del Serbal, 1984.
Macleod, R. "De visita a la 'moving metropolis': Reflexiones sobre la arquitectura de la ciencia imperial." In *Historia de las ciencias: Nuevas tendencias,* A. Lafuente and J. J. Saldaña (eds.). Madrid: Consejo Superior de Investigaciones Científicas, 1987.
Moreno, R. "Mexico." In *The Comparative Reception of Darwinism,* T. F. Glick (ed.). Austin: University of Texas Press, 1974.
Moreno Fraginals, M. *El ingenio: Complejo económico social cubano del azúcar.* 2 vols. Havana: Editorial Ciencias Sociales, 1978.
Motoyama, S. "Algumas reflexões sobre a historiographia contemporánea da ciência." *Revista de História* 103 (1975).
———. "Un análisis de la historia de la ciencia en el contexto latinoamericano." In *Historia de las ciencias: Nuevas tendencias,* A. Lafuente and J. J. Saldaña (eds.). Madrid: Consejo Superior de Investigaciones Científicas, 1987.
———. "A física no Brasil." In *História das ciências no Brasil,* M. Guimarães and S. Motoyama (eds.). São Paulo: EDUSP, 1979.
———. "História da ciência no Brasil: Apontamentos para uma análise crítica." *Quipu* 5, no. 2 (1988): 172.
Numbers, R. (ed.). *Medicine in the New World: New Spain, New France, and New England.* Knoxville: University of Tennessee Press, 1987.

Parra, P. "La ciencia en México." In *México, su evolución social,* F. Covarrubias et al. (eds.). Tomo 1, vol. 2. Mexico City, 1903.

Peset, J. L. (ed.). *La ciencia moderna y el Nuevo Mundo.* Madrid: CSIC-SLHCT, 1985.

———. *Ciencia y libertad: El papel del científico ante la independencia americana.* Cuadernos Galileo de Historia de la Ciencia 7. Madrid: CSIC, 1987.

Petitjean, P.; C. Jami; and A. M. Moulin (eds.). *Science and Empires: Historical Studies about Scientific Development and European Expansion.* Boston Studies in the Philosophy of Science 136. Boston: Kluwer Academic Publishers, 1992.

Pruna, P. M. "La recepción de las ideas de Darwin en Cuba, durante el siglo XIX." *Quipu* 1, no. 3 (1984): 369–390.

Pyenson, L. "Functionaries and Seekers in Latin America: Missionary Diffusion of the Exact Sciences, 1850–1930." *Quipu* 4, no. 1 (1986): 7–32.

———. "*In Partibus Infidelium*: Imperialist Rivalries and Exact Sciences in Early Twentieth-century Argentina." *Quipu* 1, no. 2 (1984): 253–303.

Ramos, M. P. "La difusión de la mecánica newtoniana en la Nueva España." Master's thesis, Universidad Nacional Autónoma de México, 1991.

Reingold, N. *Science, American Style.* New Brunswick, N.J.: Rutgers University Press, 1991.

———(ed.). *Science in Nineteenth-century America: A Documentary History.* Chicago: University of Chicago Press, 1964.

Reingold, N., and M. Rothenberg (eds.). *Scientific Colonialism: A Cross-cultural Comparison.* Washington, D.C.: Smithsonian Institution Press, 1987.

Roche, M. *Rafael Rangel: Ciencia y política en la Venezuela de principios de siglo.* Caracas: Monte Ávila, 1978.

Russo, F. *Nature et méthode de l'histoire des sciences.* Paris: Libr. Scientifique et Technique A. Blanchard, 1983.

Sábato, J. (ed.). *El pensamiento latinoamericano en la problemática ciencia-tecnología-desarrollo-dependencia.* Buenos Aires: Paidós, 1975.

Safford, F. *The Ideal of the Practical: Colombia's Struggle to Form a Technical Elite.* Austin: University of Texas Press, 1976.

Sagasti, F. "Esbozo histórico de la ciencia en América Latina." In *Ciencia y tecnología en Colombia,* F. Chaparro and F. Sagasti (eds.). Bogotá: Instituto Colombiano de Cultura, 1978.

———. "Underdevelopment, Science and Technology: The Point of View of the Underdeveloped Countries." *Science Studies* 3 (1973): 47–59.

Sala Catalá, J. "La ciencia iberoamericana, entre su historia y su filosofía." In *El perfil de la ciencia en América,* J. J. Saldaña (ed.). Cuadernos de Quipu 2. Mexico City: Sociedad Latinoamericana de Historia de las Ciencias y la Tecnología, 1987.

Saldaña, J. J. "Acerca de la historia de la ciencia nacional." In *Los orígenes de la ciencia nacional,* J. J. Saldaña (ed.). Cuadernos de Quipu 4. Mexico City, 1992.

———. "La ciencia y el Leviatán mexicano." *Actas de la Sociedad Mexicana de Historia de la Ciencias y la Tecnología* 1 (1998): 37–52.

———. (ed.). *La ciencia y la tecnología para el futuro de América Latina*. Mexico City: Consejo Consultivo de Ciencias-UNESCO, 1992.

———. (ed.). *The Cross Cultural Transmission of Natural Knowledge and Its Social Implications: Latin America*. Vol. 5. Acts of the XVII International Congress of History of Science, Berkeley, California, 1985. Cuadernos de Quipu 2. Mexico City: Sociedad Latinoamericana de Historia de las Ciencias y la Tecnología, 1987.

———. "The Failed Search for 'Useful Knowledge': Enlightened Scientific and Technological Policies in New Spain." In *The Cross Cultural Transmission of Natural Knowledge and Its Social Implications: Latin America*, J. J. Saldaña (ed.). Vol. 5. Acts of the XVII International Congress of History of Science, Berkeley, California, 1985. Cuadernos de Quipu 2. Mexico City: Sociedad Latinoamericana de Historia de las Ciencias y la Tecnología, 1987.

———. (ed.). "Fases principales en la evolución de la historia de las ciencias." In *Introducción a la teoría de la historia de las ciencias*, J. J. Saldaña (ed.). 2nd ed. Mexico City: Universidad Nacional Autónoma de México, 1989.

———. "Historia de la ciencia en México—E. Trabulse." *Quipu* 1, no. 2 (1984): 315–319.

———. (ed.). *Introducción a la teoría de la historia de las ciencias*. 2nd ed. Mexico City: Universidad Nacional Autónoma de México, 1989.

———. "Marcos conceptuales de la historia de las ciencias en América Latina: Positivismo y economicismo." In *El perfil de la ciencia en América*, J. J. Saldaña (ed.). Cuadernos de Quipu 2. Mexico City: Sociedad Latinoamericana de Historia de las Ciencias y la Tecnología, 1987.

———. "Nacionalismo y ciencia ilustrada en América." In *Ciencia técnica y estado en la España ilustrada*. Madrid: Ministerio de Educación y Ciencia, 1990.

———. (ed.). *El perfil de la ciencia en América*. Cuadernos de Quipu 2. Mexico City: Sociedad Latinoamericana de Historia de las Ciencias y la Tecnología, 1987.

———. "Science et développement: Une politique scientifique peut-elle tirer un enseignement de l'histoire des sciences?" In *Science and Empires: Historical Studies about Scientific Development and European Expansion*, P. Petitjean, C. Jami, and A. M. Moulin (eds.). Boston Studies in the Philosophy of Science 136. Boston: Kluwer Academic Publishers, 1992.

Saldaña, J. J.; N. Stepan; F. Safford; and T. F. Glick. "Debate." *Quipu* 2 and 3 (1985 and 1986).

Schwartzman, S. *Formação da comunidade científica no Brasil*. Rio de Janeiro: Editora Nacional, 1979.

Stepan, N. *The Beginnings of Brazilian Science: Oswaldo Cruz, Medical Research and Policy, 1890–1920*. New York: Columbia University Press, 1976.

Trabulse, E. *Ciencia y religión en el siglo XVII*. Mexico City: El Colegio de México, 1974.

———. "Introduction." In *Historia de la ciencia en México*. vol. 1. Mexico City: FCE, 1983.

Vessuri, H. "Los papeles culturales de la ciencia en los países subdesarrollados." In *El perfil de la ciencia en América*, J. J. Saldaña (ed.). Cuadernos de Quipu 2.

Mexico City: Sociedad Latinoamericana de Historia de las Ciencias y la Tecnología, 1987.

———. (comp.). *Las instituciones científicas en la historia de la ciencia en Venezuela*. Caracas: Fondo Editorial Acta Científica Venezolana, 1987.

Yepes, E. (ed.). *Estudios de historia de la ciencia en el Perú*. 2 vols. Lima: Editorial Agraria and CONCYTEC, 1986.

Zamudio, G. "Institucionalización de la enseñanza y la investigación científica botánica en México (1787–1821)." Master's thesis, Universidad Nacional Autónoma de México, 1991.

CHAPTER 1

Natural History and Herbal Medicine in Sixteenth-century America

XAVIER LOZOYA

Introduction

From the moment the territory known today as Latin America became part of the "Western world" some five hundred years ago, two different viewpoints of the same process have existed—two ways of analyzing the history of a region that for five centuries has tried to find the syncretism necessary to eradicate the great social and economic differences created by two different world views still current among the population. Latin Americans throughout these five centuries have lived and taken an active role in the conflict entailed in preserving a culture under the permanent influence and tutelage of another that has been imposed and is in power.*

On the Latin American continent, a way of relating to nature, of cutting and eating food, of understanding and curing disease, of sitting down and measuring time persists in the heirs of a world that, on Wednesday, October 17, 1492, Christopher Columbus called "Indian." This world is embodied by the indigenous inhabitants of America, from the coasts of Alaska, at the extreme northwest of the continent, to Patagonia. Columbus's geographical confusion as well as his peculiar knowledge of the Spanish language perhaps originated the widespread custom of using the terms "indigenous" and "Indian" as synonyms in America. These indigenous peoples—transculturated, acculturated, or assimilated—however one may wish to see them today, nevertheless and without true access to the printing press, still in essence preserve and transmit their knowledge orally. What Latin American city dwellers know or have known about the ideas and customs of the "Indians" we have learned from books written about them by the dominant Western culture, the owner of language and ink.

The history of science in the Americas does not wander far from this complicated circumstance. Therefore, if we are to talk about the history of science in these territories, we must begin by pointing out that there are two versions. The first follows, step by step, the history of scientific events in dominant Western cultures, primarily the European, to then research and prove that here, "on the other side of the mirror," also existed wise and studious men who drank from the same fountain, were familiar with the same notes, or were inspired by the same muse; they were, however, always solitary, surrounded by heathens, idolaters, ignorant men, uncultured men, or empiricists, according to the terminology of the age. From this perspective, the history of Latin American science in particular becomes a description of examples that attempt to demonstrate that in Indian America there was always someone who read Pliny, Buffon, Linnaeus, Darwin, or Pasteur and did it correctly. Proponents of this viewpoint then conclude that, thanks to these examples, Latin Americans have been present in the development of "universal knowledge." This is a way of interpreting the history of science inherited from Enlightenment thought in the eighteenth century that makes us believe we are moderns at the end of the twentieth.

The other version is yet to be written. It will be a version that does not judge the value of knowledge acquired by any culture, be it foreign or one's own. It is a version that, obviously, begins by rescuing and disseminating indigenous cultural baggage and believes the findings of indigenous Latin American thought, both current and ancient, to be scientific knowledge. This is in spite of the grimaces and disgust of the members of our national academies who do not realize that intellectual colonialism has outlived liberation movements and that science is also subject to fashion.

It is a paradox that, at the dawn of the twenty-first century, the world view of the dominant cultures—clearly manifested in the centers of industrial and commercial power and superficially assimilated in their colonies (today known euphemistically as "geographical areas of influence")—is shifting radically, and interestingly, with regard to the scientific knowledge of the indigenous cultures of the Asian, African, and American continents. Analyzed from a peculiar viewpoint called "ethnohistory," indigenous culture, at the end of the twentieth century, began again to scrutinize the lives of the "primitives" with the same interest as in the sixteenth century. Perhaps this is because we live in a time of transculturation similar to the transculturation during the time the New World was being explored. Even though every day the process of world integration advertised by the newspapers is discussed, what proliferates most are separatist, nationalist, regionalist, localist, and groupist movements. More and more, they recall the slow configuration of a worldwide medieval social model, formed by rich and technified "nation-city-citadels," surrounded by an infinite number of "country-villages" in total poverty. In this world of the immediate future—the "integration" of which rests on a society notably stratified and with colossal economic differences—we are witnessing the return of the use of medicinal plants and the natural resources of "primitive" cultures as a focus of interest and commerce for the hegemonic centers of knowledge and power. Just as five centuries ago, today there are new "spices" from the Orient that are already modifying Western customs, diet, and dress.

Witchcraft and magic, the para- and supranormal, create the myths that return to form part of daily life and that are reaffirmed in the taste of postmodern societies. It is the same exorcist, but now invested with new technological utensils with which to look for well-known treasures and spread the same fears.

The warriors of supertechnology have become present-day crusaders who annihilate the diabolical fanatics of formerly holy places, those places that possess natural resources necessary for Christianity. At the same time, a new millenary project entitled "God's Planet" envelops us with an energetic call to end, once and for all, the evangelizing of human beings, a task interrupted for a few brief and chaotic years by material rationalism.

These are images that explorers and naturalists of the sixteenth century knew perfectly well. They are images under whose inspiration they wrote one of the versions of the natural history of the New World to which I will refer later. The other version, that of the indigenous peoples, is barely a sketch, a brushstroke that I shall attempt to tear from the five-color palette into which the ancient Nahuas divided the world.

The study of Latin American natural history and the way in which two cultures in conflict dealt with it in the sixteenth century are enough material for a book. The following is an essay, divided into three periods, that attempts a comparative analysis of the way in which these two cultures saw nature in the sixteenth century.

First Period

> It is an owl of mortal and abominable meaning, mostly in public omens. It lives in deserts that no one can reach—a night monster with no other voice than a moan. When it is seen in cities or in the day it is of cruel and mortal meaning.
> PLINY THE ELDER

> Chicuatli or owl. It is seen as an augur when it sings above someone's house. It is said that it can signify someone's death—someone will catch a disease. It is said because the owl is the messenger of the Lord of the World of the Dead and the Lady of the Dead.
> *AUGURIOS Y ABUSIONES*

To analyze the ideas that the dominant culture of the sixteenth century had about Latin American natural history is to refer to the books that the Spaniards left behind, about the nature of the territory to which Columbus arrived in the final years of the fifteenth century. Latin American peoples of the following centuries inherited the conquerors' views of the geography, flora, and fauna of a territory in which refined civilizations had developed but whose message was interrupted. It is well known that during the first decades of colonial life, the information that the inhabitants of the Mexica Empire, for example, preserved in codices (*amoxtl*), paintings (*tlacuiloll*), or stelae, that could have testified to their way of understanding the nature that surrounded them were systematically destroyed, mostly by religious representatives of the new culture in charge of introducing Christianity. It is also true that a few missionary friars left behind testimony of the indigenous world view, collecting in their writings some of the ideas and customs of the Latin American peoples. Nevertheless, such research mainly obeyed a linguistic strategy that sought to smooth the way for the Christianization of those peoples by providing the guild of evangelizers with elements to disarticulate the philosophical foundations of indigenous religions. Thus, the information was never made public.

The results of these first true ethnographic studies, as done by Father Bernardino de Sahagún, for example, and that include entire books on the indigenous knowledge of local flora and fauna, were considered, at the time, confidential documents to which very few had access. The information was subject to such ecclesiastical censorship and control that over four hundred years had to pass before they were widely known.

Other examples are the famous indigenous herbarium, created in Mexico in 1552, the *Libellus de Medicinalibus Indorum Herbis* (Little Book of Medicinal Herbs of the Indians), and the misnamed Badiano Codex, attributed to the Indians Martín de la Cruz and Juan Badiano. If the herbarium's creation obeyed other purposes (i.e., the friars' interest in exalting the native Mexicans' knowledge of herbal medicine, thereby gaining more support from the Spanish monarch for the Colegio de la Santa Cruz de Tlaltelolco's activities), it nevertheless remained secretly archived in the Vatican's library during the same four hundred years. Its existence was not known until the twentieth century. As for the Codex, the originality of its content with regard to precolonial indigenous medical thought is highly debatable. Spanish influence is evident in the diagnosis and cure of many of the diseases and procedures described therein. It is an example of the acculturation process that was taking place in the education of indigenous peoples and, again, of Christian ecclesiastical censorship, which saw evidence of idolatry in many of the indigenous herbalist practices.

The sad truth is that we currently do not have precolonial historical sources on the subject of medicine and that reconstruction of original indigenous Mesoamerican thought (slowly taking place since the end of the nineteenth century) should be founded on the difficult task of deciphering original indigenous content in the few written sources of the first decades of the colony, thus freeing it from Spanish influence.

The sixteenth century was the time of the Spanish Empire, a time of conquest and colonization of the American continent. But it was also a time for defining and replacing many of the fundamental ideas of natural history and Western medicine that, through the interpretation of Spanish culture, had left a predominantly medieval heritage in America. This circumstance is interesting because, as we shall see shortly, in the previous century in Europe, a revaluation of classical thought, bequeathed by ancient civilizations, particularly the Greek and the Latin, had taken place. It was a predominantly Italian Renaissance that would later affect the Spanish peninsula. Sixteenth-century Spanish navigators, explorers, conquistadors, naturalists, doctors, and priests who would describe America were influenced by a view of the world and nature that was more medieval than Renaissance. Once contact with American civilizations was made and once the colonization process began, the Spaniards decided to destroy every trace of the conquered peoples' religious, philosophical, and scientific information. Their desire for enrichment and search for wealth in the new territories was always associated with the attainment of profoundly fanatical religious ideals. In the New World, the Spaniards saw the best execution of their conquest/conversion concept, a concept that they had put into practice, with less success, in Spanish territory during the Reconquest.

Now, more than four centuries later, despite its being explained as the consequence of the furor that the Christian faith had instilled in the conquistadors, the absurdity of, for example, the destruction of the Aztec city Tenochtitlan is still incomprehensible. Mercenary soldiers and fanatical priests, convinced that they were carrying out the last crusade of Christianity, killed the "Moors" and razed the "mosques" of the New World in a particularly bloody and culturally destructive campaign. The Reconquest of the Christian territories, in contrast, did not mean the Arabs' cultural destruction, and their religious antagonism—there seen as infidelity and not idolatry, as was the case in Latin America—did not lead to the destruction of all their informational and cultural sources. In Latin America, on the contrary, Christianization had as a consequence the total renunciation of the knowledge that had been accumulated for millennia as an original way of understanding the world and human existence. This world view was slowly reduced by extortion and macabre repression. Only a few of the customs and traditions practiced in the refuges of native Latin Americans survived, and this was solely through oral transmission.

It has taken five hundred years to eliminate indigenous customs and traditions, even though they are devoid of political representation and the indigenous have no access to the printing press or other formal means of transmitting knowledge. What are known today as indigenous cultures still resist culturally in the affective depths of the countries where they survive, however.

Another characteristic trait of the medieval mind-set that reached the New World can be found in the views the conquistadors had of nature. From the moment Columbus arrived, the natural world of the "Indies" fascinated sailors and explorers. On returning to Europe, these men transmitted what they had seen everywhere they could, whether in taverns or palace halls. Returning from the West Indies was a feat to talk about for the rest of their lives. Coastal lushness, the curious plants, the strange and repulsive animals—all this gave the storyteller fame and brilliance. The rabble crowded the docks when ships arrived, and brave sailors descended, bringing fantastically colored birds, unknown aromas and goods, and moribund Indians dragged into a new ecology. In the palaces, other men discussed the news and data, discussed the value of enterprises and the richness of findings. This seductive activity lasted more than a century, and many decided to leave written testimony of what they had heard or seen.

The works written throughout the sixteenth century on the natural history of the New World provided detailed evidence of basic mythical subjects that medieval thought had preserved for a long time. In antiquity, famous authors such as Pliny the Elder, Suetonius, Pomponius Mela, Saint Isidore of Seville, Bernardo Silvestre, and the Encyclopedists and travelers of the High Middle Ages had described places where there were giants; thousand-headed monsters; Cyclopes; Amazons; paradisiacal islands with milk fountains, bread trees, and medicinal plants that cured all diseases; fountains of youth; and infallible antidotes that discoveries in America made tangibly real. Beings and places were rediscovered that in ancient times had been seen as a natural part of the surroundings, and the search for them had been the stuff of daily conversation for many centuries.

Thus, for example, Ponce de León discovered, among the Carib and Lucayo Indians of America, the belief that the water of a foreign river returned vigor to the old. This legend coincided with a medieval story—that of the "Fountain of Youth," the description of which was attributed to the king of Ethiopia in 1165, called in Europe "Prester John." Archbishop Christian of Maguncia had used the document that spoke of this legendary place to recruit soldiers for the Third Crusade. If the soldiers found the "Fountain" in the country they were to be sent to, they needed only to bathe in the waters to recover the strength of youth. This had happened to Alexander the Great's soldiers, who, it was said, became strong young soldiers after standing under the stream of water. In 1513, Ponce de León thought he could find this place in the Florida peninsula. Even though the conquistador was unsuccessful, there was no lack of books that kept telling the story in the following century, but now with anecdotes about aged American Indians who became young after diving into a mighty river in Florida.

Other legendary sites described in medieval chronicles, like the Seven Cities of Cíbola, were searched for in desert areas to the north of New Spain as soon as the Spaniards heard a local story describing the existence of seven caves from which it was said that the tribes of Anáhuac had sprung. The colonizers and conquistadors from the north of Mesoamerica, gone in search of these mythological places, promised in their tales and chronicles that an infinite number of monstrous humans existed in those foreign lands, and they were none other than the Cyclopes, monopodes, giants, and pygmies of medieval tales. All of the authors of the time touched on the theme of extraordinary beings and fantastical places. Let us look at a few more examples.

Joseph de Acosta wrote in *Historia natural y moral de las Indias* (Natural and Moral History of the Indies, 1590) that he had found evidence in Peru of colossuses that, according to medieval history, had lived in ancient times. There were traces in the New World, close to the mercury mines where, according to Acosta, there existed "a fountain running with hot water and as the water runs it turns to stone." There the author claimed to have seen "trees one part of which give fruit one half of the year, and the other the next half." He also said that he heard of "bees so small their hive is underground," of "fish that as they leave the water turn to butterflies," and of a wonderful plant called coca that "one only need carry it in the mouth, that he may walk twice as far without eating anything else"—a plant that still fascinates present-day warriors and gladiators.

Gonzalo Fernández de Oviedo included in *Historia general y natural de las Indias* (General and Natural History of the Indies, 1526), a section dedicated to fantastic animals of the New World—a section that has the characteristics of a typical medieval bestiary. He describes the "horrendous four-footed snake" called iguana, "dragons" like those in countless medieval tales, chameleons walking on coals, and giant serpents, such as the anaconda, that with a squeeze could tear a man apart. These creatures of the American tropics, unknown to Europeans, confirmed the stories and drawings of ancient naturalists, or at least complemented them. The Europeans' horror before such chilling creatures is understandable, as is the amazement they convey in their descriptions of "fantastical" beings that still amaze us to this day, such as "butterfly-birds" (hummingbirds), ants that devour trees in one night, and scorpions the size of a man's palm.

Pedro Mártir de Anglería, in his *Décadas del Nuevo Mundo* (Decades of the New World, 1493–1510), also frequently dealt with natural things in recently acquired territories and enjoyed creating horrific descriptions, such as his description of the celebrated Mexican marsupial, the opossum: "Among said trees was found a monstrous animal, with the face of a fox, tail of a Cercopithecus, ears of a bat, human hands, and feet of a monkey; that wherever she goes, carries her children on the outside of her belly like a large bag." A description like this must have dropped the jaw of any European reader.

Based on their reading of books called "classics" of antiquity, such as Pliny the Elder's *Natural History*, written around AD 70, the naturalists of the sixteenth century saw in every step of their exploration of the New World the corroboration of references to beings, places, and plants from a territory the dimensions of which varied daily and with characteristics as aberrant as a rainy season in summer.

Pliny wrote the following at the beginnings of the Christian era: "Crates Pergamenus calls certain Indians [*sic*] that live more than 100 years *gimnetas*, even though many others call them *macrobios*. Clesias says that certain of these so-called *pandores* live 200 years with white hair in youth and dark hair when they are old. Their mates give birth only once, because they subsist only on locusts." European scientists of the sixteenth century were interested in locating places that contained specimens like those described by Pliny. Francisco Hernández, the famous royal physician to Felipe II, explored America from 1571 to 1577 (where he wrote *Historia natural de Nueva España* (Natural History of New Spain). In his translation of and commentary for Pliny's work, he responds, "I have seen these locusts to which the author refers, and the ones that eat them are the *Indians* of New Spain. They are fried and taste good, similar to sweet almonds." Surely, he is referring to some of the varieties of insects that the Mexicas considered edible and some of which, like stink bugs, still strike fear into the hearts of some tourists in Mexico's markets.

Regarding the confirmation of the existence of long-lived peoples, Juan de Cárdenas, a doctor in New Spain and author of *Problemas y secretos maravillosos de las Indias* (Problems and Secrets of the Indies, 1591), dedicates some chapters to explaining why the hair of the Indians of New Spain never turned white, why they did not grow beards or go bald, and why they lived longer than anyone else. The natural cause of this wonder was, according to the author, the same that made trees in the West Indies grow roots above the earth and never lose their leaves and that made for a wheat, corn, and fruit harvest all year round: in short, the variety of earth's "tempers" and "humors" in human beings that reacted to these lush, prodigious, and fertile lands, which, "if only Pliny could see them he would be awestruck and frightened." With his gaze set firmly on ancient readings, Cárdenas warned us of Latin American nature:

Who heard to be tried and true—
... that the Eagle's stone tied to the thigh tears the child from the womb,
... that the carbuncle amid darkness should give light and shine,
... that the plant called Baaras by the Indians casts out demons and he who pulls it first, be animal or man, will die,
... that the animal called hyena with only its shadow puts all other animals to sleep,
that the fish called Remora can stop a ship by only coming near it,
... that the Unicorn's horn, when put before any poison, will sweat,
... and another thousand strange properties, that the ancient writers tell us of—will understand why I cannot cease to give credit to the wonders and occult secrets that with so much testimony of truth we may write of this New World of the Indies.

While I have said that the main desire of the explorers was to confirm what was written in antiquity in the classics rather than carrying out research on the characteristics of the New World, there were also moments in which some of authors, showing a critical eye, did not hesitate to correct the ancients as they described their own experiences, thus emphasizing the novelty of these newly discovered lands. Let us look at another example comparing Pliny and Hernández.

Pliny wrote:

> The *Arimaspi* are men singled out by the one eye they have in the middle of their forehead [and] are perpetually fighting with the *griffins*, a type of fierce bird that takes gold from certain mines to protect it from the *Arimaspi* [who] with the same greed steal it from the *Scythians* who eat meat and are called *anthropophagous* by the Greeks. In a great valley surrounded by Mount Imaus there is a region called Abarimon where there are wild men with their feet turned backwards who live among the wild animals. These die upon removing them from their region and therefore cannot be sent alive to kings nor could they be taken by Alexander the Great. (My translation; original emphasis)

Francisco Hernández believed he had found the explanation to Pliny's text in his knowledge of America:

> There are apparently only two *Scythias*, one in Europe that borders Sarmatia, also called Poland, and another Asian one near Mount Imaus. Of the inhabitants of one and the other, Pliny said that there are not a few who eat human flesh. It is not only they, and they do not eat only human flesh. Among the Ethiopians, whose king has one eye in his forehead, there were *agriophages* who lived mostly off lion's meat and the *pamphages* off birds and the *anthropophagites* off human flesh. In our time, in many parts of the West Indies, they used mainly sacrificed men, of whom, by the way, many were Spaniards. (Original emphasis)

And he adds:

> The *Arimaspi* are people who live near the Riphean mountains together with the *Hyperboreans*, *Essedones*, *Arimphaei*, and *Cimerians*, even though it is understood that these did not have the single eye, nor were the *griffins*, which Pliny mentions, those birds. While I was writing this, a soldier from Peru told me that in the Province of Homagua in the West Indies there are birds so large that they snatch the Indians away with their talons and eat them and tear them apart in the air and that they cover, with their wings, the gold that is in great abundance in these lands and above make nests and breed. A feather from their wings is as thick as an arm. I would not be able to determine if these are the *griffins* or vultures or some other kind of bird, nor do I mean anyone else to believe it, but it seemed to me worthy of a warning among the many other monstrosities that we can see every day in these lands. (Original emphasis)

The fantastical griffin turned out to be insignificant next to the Andean condor, which, every time it flies by with a cow in its talons, feeds the legend and makes humanity's time shrink.

Second Period

What can we say about the view the indigenous peoples of Latin America had of nature and some of its animals in the sixteenth century? Having had a profoundly religious world view, the inhabitants of Mesoamerica believed the creatures in nature shared the cosmos with them. In Nahuatl culture, the thirteen skies and nine levels of the underworld were inhabited by a multitude of gods and beings that, day and night, separated into diverse mythological categories. There was no dividing line between man and other animals, and, unlike in Christian thought, in the indigenous Latin American world, human beings were not at the center of nature. The surface of the earth was understood to be a platform divided into four segments, or petals, joined at the center by a navel and surrounded by a sea that rose at its ends to form walls capable of holding up the sky. The sky was, in turn, composed of four lower floors and nine upper ones. Each of the great earthly flower's segments had a color corresponding to its position in space: black, white, blue, and red corresponded to the cardinal points. The center of the flower was green and shown in precious jade. Alfredo López Austin writes in *Cuerpo humano e ideología: Las concepciones de los antiguos nahuas* (Human Body and Ideology: Ancient Nahua Viewpoints, 1980) that the world of the Nahuas was not inhabited only by humans, plants, and animals. Other, less-perceptible, beings surrounded humans and influenced their life—beneficially or otherwise. Some were divine characters that, in their own times and zones, visited the surface of the earth or the lower layers of the sky or lived permanently between them; others were their envoys, their messengers; still others were guardians or owners, custodians of animals, vegetables, fountains, streams, or hollows; others were locked in rocks or threatened humankind with their terrible descent from the sky. In any case, it was a world complexly populated that was explained by its own creation myths.

Generally speaking, for most native Latin American cultures, there were two major categories of animals. The first were those that belonged to the complex sphere of myth, predominantly religious, but they were also actors in innumerable omens and fantastical beliefs, beings that were invisible or seen rarely by a privileged few, which gave credence and strength to the legend of which they were a part. The second category was those found in the quotidian simplicity of existence without, however, ceasing to be inserted—when required—into the magical world that was an indivisible part of the existential reality of those peoples. The Indians described in Nahuatl, with great patience and care, examples of these categories to Father Bernardino de Sahagún, who translated the information into sixteenth-century Spanish. The following stories come from the *Historia general de las cosas de Nueva España*:

> The tiger moves and makes a racket between rocks and crags and also in the water, and they say it is the prince and lord of other animals and that it is sagacious, cautious, and presents itself like a cat and that it feels no physical strain and is disgusted with the thought of drinking filthy, stinky things; it takes great care of itself, bathes, and by night watches the animals it will hunt, it has excellent long sight, even when it is dark, and even if there is fog it can see very small things. When it sees the hunter with his bow and arrows, it does not run but sits, watching him, without getting behind anything, or close to anything, and then it begins to hiccup, and that air goes straight to the hunter so as to strike fear in him and make his heart faint with the hiccup, and the hunter then begins to shoot at it, and the first arrow, which is made of cane, the tiger takes in its paw and tears to pieces with its teeth and begins to scold and growl . . .
>
> Skillful hunters, on firing the first arrow, if the tiger tore it to pieces, take a leaf from a tree, thrust it onto the arrow, and shoot it at the tiger. With the leaf placed in this way, it makes a noise like that when the locust flies, and it falls on the ground halfway to, or near, the tiger, and with this it [the tiger] is distracted as the arrow arrives and passes by it or injures it, and then the tiger jumps up and falls to the earth and then sits as it was before, only to die sitting, without shutting its eyes, and even though it is dead, it seems alive.
>
> An assassinlike people, those called *nonotzalique* were daring people, used to killing, they carried the skin of a tiger with them, with one piece on their forehead and another on their chest, and they carried the tail and claws and heart and fangs and muzzle. They said that with this they were strong and frightened everyone, and everyone feared them.

One of the more surprising creatures of fantastic indigenous fauna was the river otter, whose description competes in strength and mystery with the mythologies of the ages:

> There is an animal that lives in the water called *ahuizotl*; it is the size of a small dog, has short, fine hair, very small and pointy ears, a black,

smooth body, it has a long tail and at its end a sort of human hand. It has hands and feet that are like a monkey's; this animal lives in the deepwater springs; and if anyone reaches the edge of its habitat, it snatches them with its tail-hand and brings them underwater and takes them to the depths and then disturbs the water and raises waves; it seems as if there is a tempest and the waves break on the shore and make foam and then out come many fish and frogs from the water's depths and they walk over the water's surface and make a great disturbance. And the one that was dragged down there dies, and in a few days the water throws up his body and he appears without eyes and teeth and nails, which are what the *ahuizotl* takes from him.

The sixteenth century left us a formidable indigenous account of Latin American snakes, the beauty, strength, and danger of which are still written about today: "There is a snake with two heads: one where the head should be and the other where the tail should be, and it is called *maquizcoatl*; it has two heads and in each one has eyes, mouth, and teeth and tongue, with no tail. It is not large, or long, but small and has four black stripes on its back and another four red ones on one side and another four yellow ones on the other. They call gossips by the name of this snake, as they say they have two tongues and two heads."

The legends of animal origins are numerous in the indigenous world. For example, the *ajolote*, or *axolotl*, the classical tiger salamander of Central Mexican lakes, was considered to be one of the incarnations of Xolotl, the jesting deity of transformations, and of those disguised. The legend says that the other gods once chased the happy, eternally changing character:

> After the sun and moon rose above the earth, they were still, not moving from one place, and the gods said: "How can we live? Does the sun not move?"
> "Must we live among peasants?"
> "Let us all die and make them resuscitate on account of our deaths!"
> And then Air killed all the gods and it was said that one called Xolotl refused death and said to the gods: "Oh gods, I do not want to die!" And he cried greatly until his eyes grew swollen, and when the one who killed reached him, he ran and hid among the cornfields and became a cornstalk, the one with two stalks that the farmers call *Xolotl*; but he was seen and found among the cornstalks; again he ran and hid among the magueys and became a maguey with two bodies, the one called *mexolotl*, but again he was seen and began to run and

jumped into the water and became a fish, the one called *axolotl*, and from there he was taken and killed.

At the fire god Xiuhtecutli's party, children's fathers and mothers hunted: some snakes, others frogs, others fish called *xouiles*, and others playful animals called *axolotl*, and they threw these on a grill at home, and once they were toasted the children ate them and said, "Our father fire eats toasted things!"

Francisco Hernández's description of this little animal contrasts with the Mexicas' poetic vision and perhaps explains why in the following centuries Spanish medical books said that "ajolote syrup" had aphrodisiacal properties:

The *axolotl*, or *juego de agua* is a type of lake fish covered in soft skin with four lizardlike feet. It has a vulva similar to a woman's, and its belly has dark spots; for a tongue, it has short, wide cartilage; it swims with its four feet, which end in fingers similar to a frog's: its head is sunken. It has been repeatedly observed that it has menstrual flows as women do, and when it is eaten it excites genetic activity, just as the *estincos*, which some call land crocodiles and are perhaps of the same species, do. It provides healthful, tasty food, the Spanish season it with clove, the Mexicans with peppers [chiles] which they greatly enjoy ground or whole. It takes its name from its enjoyable form.

The *huitzitzilin*, "the buzzing one," the hummingbird, is one of the birds with the greatest legends in tropical America. Its prodigious flight and tiny size captured the natives' imagination. They considered it to be a mythological being par excellence. Sahagún was told of these birds as follows:

They are very small and seem more like horseflies than birds, and there are many kinds; they have a small beak, black and small like a needle. They buzz in the bushes, there they lay their eggs and brood until they hatch. They lay no more than two eggs. They eat or gain sustenance from dew on flowers, as bees do. It is very light and flies like an arrow. It is reborn every year; in winter it hangs from trees by the beak; hanging there, they dry and their feathers fall out; when the tree turns green again, it awakens, begins its rebirth, and when the thunder falls for rain, it flies and is born again. They are medicine for pustules, and he who eats them will never have pustules, but they will make the one who eats them sterile.

The great variety of species that existed in Mexico were carefully classified, serving as an example for some Nahuatl zoological nomenclature:

> Of these little birds, the ones with colored throats are called *quetzal huitzilin*, with bright red elbows on their wings, a green chest and their wings and tail look like fine quetzals . . .
> Others are called *xi huitzilin* and are all blue . . . like resplendent turquoise . . .
> Others that are called *chalchi huitzilin* are light green like plants,
> The ones called *yiauhtic huitzilin* are purple . . .
> The ones called *tlapal huitzilin* are colored and mixed with black . . .
> The ones called *aiopal huitzilin* are light purple.
> The ones called *tle huitzilin* are resplendent like hot coals,
> The ones called *quapa huitzilin*, are a tawny yellow.

Father Diego Durán, in *Historia de las Indias de Nueva España e islas de la tierra firme* (History of the Indies of New Spain and the Islands of Terra Firma, 1581), refers to the hummingbird in almost the same terms:

> The most celebrated and solemn festivity of all this land is of the idol Huitzilopochtli . . . it was a wooden statue, carved in the shape of a man sitting on a blue wooden seat. He had a crest of plumes on his head in the shape of a bird's beak, the one they call "*huitzitzilin*," which we call "hummingbirds," which are blue- and green-feathered birds. They have a long black beak and very shiny feathers. When they feel winter coming, they move to an evergreen tree and with natural instinct look for a hollow. They sit on a branch next to the hollow and stick their beak in it as far as they can and they remain there for six months of the year—the entire winter—finding sustenance only on that tree, as if dead. When spring comes and the tree gets new growth and grows new leaves, the bird, helped by the vigor of the tree, resuscitates and goes forth and procreates. Because of this, the Indians say they die and are resurrected. I have seen it with my own eyes in winter.

Third Period

In sixteenth-century Spain, herbal medicine was at the center of medical interest, and knowledge of it was required for making "simples," used to cure illness. Frequently, the ingredients of a *pharmako* inherited from ancient

Greece took years to reach Spain from the Near East, and when they did, they arrived in small quantities and with diminished curative effects. They came from territories where these vegetable products had been used for centuries, gathered mainly from wild specimens, rarely grown, and were on the verge of extinction. Doctors invested a great deal of work in evaluating the quality of these products and in writing reiterative texts in which they gave precise instructions about how to cut, clean, and process medicinal herbs. The general opinion among Spanish doctors on either side of the ocean was that the New World was a wonder-drug gold mine, despite the incorrect uses to which the native Mexicans put the plants. Provided by God, American resources were the salvation of a purulent, arthritic, kidney-stoned Old World. These diseases and afflictions also hurt the Spanish population of the American colonies.

Colonial medicine busied itself with finding "Indian" cures and locating Latin American species that resembled known, trustworthy European plants. Health was understood to be a consequence of the equilibrium between the body's four humors: blood, yellow bile, phlegm, and black bile. The Greek *eurythmia*, which Hippocrates turned into a way of life, was, for the medieval man, a combination of internal liquids; disease could only be explained as the body's excess or lack of one of the humors. Hence, causing excessive sweating, intestinal laxation, bleeding, inducing vomiting, and evacuating phlegm were the most common healing methods. Restoring the humors lost during disease or evacuated intentionally was achieved by feeding on products clearly differentiated by their capacity to generate bile, blood, or phlegm.

American medicinal plants, depending on their physical properties, were used for some of the aforementioned procedures. The properties of a plant were determined through simple organoleptic procedures consisting of trying, smelling, or rubbing plants between the fingers, for example. Bitter plants induced vomiting; dry and warm plants were laxatives; moist and dry ones stopped diarrhea; cold and dry ones moved bile; and so on. There was an emphasis on hot flavors, on the urtication produced by juices on mucus, on the smell of latex and resin, the consistency of oils and mucilage, to, in the end, incorporate them into a medicinal classification that recognized the four categories—"cold," "hot," "dry," and "moist"—that combined with their philosophical equivalents—"phlegm, choler, melancholy, and blood." All were qualified into four degrees of intensity, for example, a plant could be "cold" in the second degree and "moist" in the first. The exceptions to these classifications, however, and the existence of intermediate categories ("temperate," e.g.), would, frequently, be more numerous than the rules themselves, but the Aristotelian method of reasoning, in the end, put everything in its place. Juan de Cárdenas provides an example:

> Why, if *cacao* is cold, does *chocolate* give us the effect of great heat?
>
> First, *cacao*, without being toasted or prepared in any way, has the property of constricting the bowels, stopping menstrual periods, closing urinary tracts, obstructing the liver and especially the spleen, depriving the face of its vivid, natural color, weakening digestion, causing paroxysms and fainting, and, in women, blushing, female problems, and, above all, it engenders perpetual anxiety, melancholy, and heart murmurs.
>
> Now, on the other hand, we can see that if *cacao* is ground and toasted, even if it is not mixed with anything other than some atole, which is simple Indian food, with only this we can observe that it fattens, provides sustenance, causes urination, is a healthy remedy for obstruction, helps digestion, rouses the appetite, helps and heals female problems, causes happiness and vigor.
>
> Reasons for such a notable property. The dominant part of *cacao* is cold, dry, and thick, earthy and melancholic, and thus is the cause of this damage. The nature of a second part of *cacao* is airy, oily, warm, and damp in complexion and is the one that forms *chocolate*, which is bland, lenitive, and amorous. It has a third part, which is very hot, penetrating and belongs to fire, and this is the bitter taste that promptly rises to the brain, causing sweat, provoking periods, removing obstructions, and moving the bowel's excrement.
>
> We therefore conclude that *cacao* is composed of contrary parts; thus, the oily part (which is moist), removes some dryness from the exceedingly hot and, at the same time, although it should be cold in the second degree, it is in reality in the first, because the other parts are hot and remove one, and thus everything that remains of the *cacao* is cold in the first degree and dry in the second and composed of a substance that is thick, earthy, and melancholy. (My emphasis)

On either side of the ocean, Spanish doctors experimented with the effects of New World medicinal plants. The study of flora encapsulated two aspects: to acclimatize the new plants to Mediterranean soil; and to evaluate curative effects by using experiments concerning taste, texture, color, and other properties inferred from the plant's morphology. These new products were administered to the ill to test their effects in moving the humors and to later infer the course of this new treatment.

In most cases, acclimatizing American plant species to orchards in Spanish hospitals and monasteries failed. Therefore, the project was soon abandoned, and preference was given to importing seeds, leaves, pulverized bark, dry fruit, and so on—trade that was always threatened by problems of adulteration and fraud in the massive shipments of Latin American plant material arriving in Cádiz.

Nicolás Monardes—the doctor from Seville who dedicated a large part of his life to the study of materia medica and whose books were highly successful—is considered to have formally introduced Latin American plant-based medications in Europe:

> From New Spain they bring that liquor that for its excellent and wonderful effects is called *balsam* in imitation of the true Egyptian balsam. It is made from a tree with thin, closed leaves in the form of nettles. The Indians call it *xilo* and we call its issue balsam. Its use is purely medicinal and is ancient, almost dating to the discovery and conquest of New Spain, for the Spanish knew of it, because with it they healed the wounds they received from the Indians. When it was brought to Spain for the first time, it was kept for its wonderful effects. One ounce was worth ten and twenty ducats, and now an arroba is worth three or four ducats. This is one of the qualities of abundance or rarity: when it was expensive, everyone made use of its powers, and since it dropped in price, no one uses it at all.
>
> Even though the Indies were not discovered to find this wonderful liquor, the work of our Spaniards was put to good use, since Egyptian balsam has been gone for many years.

It used to be customary to believe that in the places where a disease was contracted its cure existed. Since syphilis riddled the Old World's population, explorers of Latin American nature searched zealously for new remedies for this disease, especially when, because of the polemics concerning its origin, it was decided that the New World was the cradle of this ill. In fact, the proof that was used to back up this theory was that on the Latin American continent there were "innumerable" resources to cure this "purulent affliction" that, it was said, ardent Indian women had transmitted to conquering armies.

The bark of a chinaberry and the root of a sarsaparilla would be the two stellar products for treating syphilis in the sixteenth century and produced a commercial fever up to then unknown. In a few years, chains of specialized cliniclike establishments appeared with the sole purpose of preparing this curative potion and treating patients with syphilis. Monardes attentively follows this process:

It was Our Lord's will that from whence came the purulent affliction should come its remedy, because pustules came from this part of the Indies and Santo Domingo.

The chinaberry was discovered after the Indies were discovered. An Indian gave notice to his master of this in the following manner: since a Spaniard was suffering great pains from the pustules he contracted from an Indian woman, the Indian, who was one of the doctors of that land, gave him chinaberry water, with which his pains not only disappeared, but with which he was also cured. Thus, many other infected Spaniards were cured, and this was communicated to us by the recently arrived in Seville, and from here it was divulged throughout Spain and from Spain to the world.

Spanish doctors' opinions were not always so generous concerning the actions of Indian doctors. Legends or stories related to important Latin American plants, however, all have the same plot: the Indian doctor, silent and docile, gives his Spanish master the solution to his, or his relatives', affliction. This occurs with Cinchona bark, *ipeca*, *raíz de Jalapa* (a type of morning glory), sarsaparilla, and many other plants. The learned doctor Francisco Hernández, in gathering medicinal flora, had his own opinion of indigenous herbal medicine and Indian doctors: "I will not speak of the perfidious confabulation of the Indians, of the perverse lies with which they deceived the unwary, speaking falsely and with trickery; nor of the many times that, trusting fallacious interpreters, I believed false virtues of plants—and I barely managed to combat their noxious effects with the medical arts and the notable favor of Christ."

Sahagún's informants would answer the Spanish royal physician's misgivings by saying,

> The doctor tends to cure and remedy diseases; a good doctor is an expert, he is knowledgeable of the properties of herbs, stones, trees, and roots, experienced with cures. He is one whose business it is to know how to arrange bones, to purge, to bleed, to make incisions and stitch them, and, in the end, to free from death's door. A bad doctor is a trickster and, because he is not skillful, instead of healing, he worsens the illness with the potions he gives them and sometimes uses witchcraft and superstition to lead others to think he provides a good cure.

Sixteenth-century Spanish medicine had its own medicinal plant-based recipes as well as countless other natural resources, the use of which, analyzed from a vantage point centuries away, shows a far-from-useful empirical base. Latin American herbal medicine, with its newly discovered plants, reinforced a medical practice that was not tremendously different from the one practiced by indigenous American peoples and that, in many cases, was superior to its European counterpart. In *Tesoro de medicinas* (Medical Treasures), Gregorio López gives some examples of Spanish medicine:

> *Dysentery* is bloody bowel movements with intestinal scrapings. Treatments:
> It is said that the remedy is to frighten the patient or to remove the heat from fresh horse manure and drink it, or to squeeze the same manure mixed with wine and drink it, or to put a dry cupping glass on the back of the neck, or to gulp watered down vinegar, or to put a stone drenched in blood in the patient's hand, or to drink dog urine with wine, or ground pig's feet with wine.
> *Broken ribs:* dry ground goat manure baked with wine is to be plastered onto the broken ribs.
> *Chastity:* Things that help one to be chaste.
> To carry verbena helps, or to eat basil, or to eat rue diminishes sperm, or to eat spearmint after a fast, or eaten purslane tempers venereal ardor, or lettuce seeds, or to eat Castille pumpkins, or to drink coral tempers the sperm, or to carry a jasper represses nocturnal pollution.

Let us now take a look at some examples of herbal treatments from Latin America:

> *Diarrhea.* For either children or adults, in the case of diarrhea, they must drink water boiled with *tzipipatli*. Three, four times they must drink, and the mother should as well, since the baby will drink the medicine from her milk. An adult must drink atole, atole with *chia* with tortilla pieces mixed in.
> The trees on which guavas grow are small with sparse leaves and branches. The fruit of these trees is called *xalxocotl* [sandy fruit]. They are very good to eat and stanch bowel movements. The leaves boiled in water are given to patients with dysentery.
> *To expel the seed:* There is another medicinal herb called *oquichpatli* [medicine for males]. Its root, ground, is advantageous to men and women who, because they have not expelled their human seed, or out of fear, or because on another occasion have been cut, start to dry up and

cough continuously. Taking the ground root expels a stinky humor in two or three days and through the member produces white urine, like stinky lime water, which also happens in women. This is also medicine for someone who has expelled the seminal humor in his dreams. With this medicine excessive pleasure or permanent desire is terminated. The quantity of this root should be half a finger, ground only once. This plant is found in the fields of Tulancingo.

Breaks. In case our spine or ribs or any bone may be harmed. First the bone must be pushed, stretched, and reset. Immediately afterward, the *zacacili* root must be cut, put in a thick poultice, and the injured part splinted. If there is swelling around it, it is pricked with an obsidian knife and *iztaczazalic* is applied on the swelling mixed in with *tememetatl* root.

Before this information was transcribed, it was examined by the Mexican doctors who had dictated it in Nahuatl, and Father Bernardino de Sahagún compiled their names. They are without a doubt the only native Mexican doctors we know whose message comes from the sixteenth century:

Juan Pérez, from San Pablo
Pedro Pérez, from San Juan
Pedro Hernández, from San Juan
José Hernández, from San Juan
Antonio Martínez, from San Juan
Miguel García, from San Sebastián
Baltasar Juárez, from San Sebastián, and
Francisco de la Cruz, from Xihuitonco.

Note

* I intend this essay to be easy to read and hope to disseminate basic ideas on this subject that I have tried to summarize. My sources are numerous, and the bibliography is long. I have therefore decided against any pretense at erudition by avoiding detailed textual references. I include only a limited bibliography. There the interested reader may corroborate any liberties I might have taken.

Bibliography

Acosta, J. *Historia natural y moral de las Indias*. Edited by E. O'Gorman. Mexico City: Fondo de Cultura Económica, 1979.

Augurios y abusiones. Textos de los informantes de Sahagún. Mexico City: Instituto de Investigaciones Históricas, Universidad Nacional Autónoma de México, 1969.

Cárdenas, J. de. *Problemas y secretos maravillosos de las Indias*. Mexico City: Academia Nacional de Medicina, 1977.

Colón, C. *Diario de a bordo*. Edited by C. Sanz. Madrid: Biblioteca Americana Vetustísima, 1962.

De la Cruz, M., and J. Badiano. *Libellus de Medicinalibus Indorum Herbis*. Mexico City: IMSS, 1964.

Durán, D. *Historia de las Indias de Nueva España e islas de la tierra firme*. 2 vols. Edited by A. M. Garibay. Mexico City: Porrúa, 1967.

Fernández de Oviedo, G. *Sumario de la natural historia de las Indias*. Edited by J. Miranda. Mexico City: FCE, 1979.

Glick, Thomas F. *The Comparative Reception of Darwinism*. Austin: University of Texas Press, 1974.

Hernández, F. *Historia natural de Nueva España*. 2 vols. Mexico City: Universidad Nacional Autónoma de México, 1959.

López, G. *Tesoro de medicinas*. Edited by F. Guerra. Madrid: Ediciones Cultura Hispánica del Instituto de Cooperación Iberoamericana, 1982.

López Austin, A. *Cuerpo humano e ideología: Las concepciones de los antiguos nahuas*. 2 vols. Mexico City: Instituto de Investigaciones Antropológicas, Universidad Nacional Autónoma de México, 1980.

———."De las enfermedades del cuerpo humano y de las medicinas contra ellas." *Estudios de Cultura Náhuatl* 8 (1969): 51–122.

———. "De las plantas medicinales y de otras cosas medicinales." *Estudios de Cultura Náhuatl* 9 (1971): 125–230.

———. "Textos acerca de las partes del cuerpo humano y de las enfermedades y medicinas en los primeros memoriales de Sahagún." *Estudios de Cultura Náhuatl* 10 (1972): 129–153.

Mártir de Anglería, P. *Décadas del Nuevo Mundo*. 2 vols. Edited by E. O'Gorman. Mexico City: Porrúa, 1964.

Monardes, N. *Herbolaria de Indias*. Edited by X. Lozoya, E. Denot, and N. Satanowsky. Mexico City: IMSS, 1990.

Plinio Cayo, S. *Historia natural*. 2 vols. Translated and annotated by F. Hernández. Mexico City: Universidad Nacional Autónoma de México, 1966.

Sahagún, B. *Códice Florentino*. 3 vols. Biblioteca Medicea Laurenziana, Italia/Archivo General de la Nación. Mexico City: Giunti Barbera Florencia, 1981.

———. *Historia general de las cosas de Nueva España*. 4 vols. Edited by A. M. Garibay. Mexico City: Porrúa, 1977.

Weckmann, L. *La herencia medieval de México*. Mexico City: 2 vols. El Colegio de México, 1984.

CHAPTER 2

Science and Public Happiness during the Latin American Enlightenment

JUAN JOSÉ SALDAÑA

Hispanic America's Historical Dynamics

The Enlightenment in the Americas was simultaneously the cause and the effect of social and cultural changes in the region. Changes increased in intensity during the eighteenth century and during the first third of the nineteenth. In that period, social and economic life in the colony became more dynamic; there was educational, cultural, and scientific secularization; and a Creole nationalist consciousness and revolutionary movements emerged in Latin America. The Enlightenment ideal came to fruition in the arts, history, literature, urbanism, ethnography, philosophy, linguistics, and, especially, science and technology.

In Spain's colonial empire in the Americas beginning in the sixteenth century and, to a lesser extent, in the Portuguese Empire, geographical distance and time brought about a certain autonomy to the colonies. In the eighteenth century, it was evident in several places in Hispanic America that societies were no longer in the process of formation or subject, therefore, solely to advances in the home country—which had been the case in the two previous centuries under absolutism.

This new autonomy may be clearly seen in the cases of the Viceroyalties of New Spain and Peru, as well as in the Viceroyalties of Río de la Plata and New Granada and in the captaincies general and governments of Guatemala, Cuba, Quito, Popayán, and so on, although in the latter, autonomy started in the second half of the century as a consequence of the widespread economic recovery which took place then. In practically every Hispanic American territory at the time, there was a really important socioeconomic dynamic and a local cultural awakening of which science was a part.

The societies in the extensive Hispanic American territory enjoyed a diversified, growing economy. Mining, agriculture, and handicrafts constituted the main activities. The domestic market relied on purchasing power created by mining and miners. Thus, mining provides us with the most important indicator of the Hispanic American colonies' prosperity. Between the fifth and the sixth decades of the eighteenth century, recovery began in the mines at Potosí, Charcas, Chocó, Popayán, and the Viceroyalty of New Spain. This recovery brought gold and silver production out of the crisis it had been in for the last century, and at the end of the eighteenth century and the beginning of the nineteenth, it helped mining reach its highest production levels during the colonial period. In Mexico, for example, twenty-seven million pesos' worth of gold and silver coins were minted in 1800, compared with four million coined in 1700.[1]

Agricultural and handicrafts production kept pace with the intense activity in the mining centers. Highland Peru was for Quito, Cuzco, Arequipa, and Buenos Aires what the north of Mexico was for the central and western Mexican lowlands: consumers there were able to stimulate domestic trade at long distances.[2] Cotton, sugar, wine, wood, firewood, straw, mate tea, coca, mules, suet, tobacco, wool, leather, textiles, and many other "products of the earth" were created by local farmers and craftspeople in active commercial circuits, which brought together widespread regions of the continent. Also, an important system of land, river, and coastal communications ensured the transportation of and trafficking in goods.[3]

Intense economic activity was needed for the development of diverse locally produced materials and intellectual inputs, because it was not always possible to obtain these from the distant and sometimes chaotic home country. Indeed, this need triggered the search for raw materials (like mercury and iron) in several places and the development of technical innovations for industry (in the extraction and processing of minerals and the minting of currency and coinage, etc.) and for agriculture (e.g., for the cultivation of sugarcane, tobacco, silk, cotton, and indigo). These innovations occasionally or permanently broke the bans formerly imposed by the home country.[4]

In order to obtain these inputs, geography and natural resources were surveyed. It was quickly understood that by developing technology, Hispanic Americans would also contribute to increasing the wealth and prosperity of their territories, no longer for Spain's or Portugal's exclusive benefit but for the colonies' benefit as well.

The participation of experts (miners, botanists, geographers, engineers) with adequate scientific and technological training, as well as the creation of modern scientific institutions that could offer the required courses of study, gradually became necessary. At the end of the eighteenth century, individual and erudite cultivation of knowledge was replaced by an interest in the "useful arts." Simultaneously, society demanded scientific and technical knowledge.

Economic-modernization initiatives frequently began with the interested parties, and always with their participation in financing and running the projects. Royal sanction was bestowed once the ideas, modi operandi, viability tests, and even financing had already been contributed by the Hispanic Americans. This was a change in the cultural attitude of the most dynamic sectors of colonial society, a change that found its inspiration in the Enlightenment.

Economic growth led to the formation of powerful and influential corporations of mine owners and merchants. In New Spain and Peru, in the 1780s, mine owners organized into bodies that had their own tribunals, and merchants organized trade groups.[5] Trade groups were created in all important Hispanic American cities at a rapid rate because of trade liberalization (in Mexico in 1594; Lima, 1618; Caracas and Guatemala, 1793; Buenos Aires and Havana, 1794; Cartagena, Chile, Guadalajara, and Veracruz, 1795).[6]

Merchants also accumulated considerable fortunes, which were invested in mines and country properties; this accounts for the prosperity of several areas in Latin America at the end of the eighteenth century. Mining fortunes also became very significant, and some mine owners—for example, José Bustamante Bustillo and Pedro Romero de Terreros in New Spain and Francisco López Calderón in Peru—besides being rich, were also entrepreneurs, promoters, and experts in guild organization.

Economic progress was followed by social development typified by the economic societies established in almost all of the Americas.[7] The most advanced sectors inside these patriotic associations, encouraged by an enlightened bourgeois mentality, met with the mine owners, merchants, and other progressive sectors from the clergy and the army to struggle for economic growth, understand the scope of the territories and their natural wealth, and work for important educational reforms. Schools were created to, for example, provide scientific and technical instruction for mine owners, metallurgists, engravers, draftsmen, engineers, architects, farmers, druggists, seamen, artists, and other artisans. In Mexico, at the urging of and with the support of mine owners and merchants, schools of mining (in 1792), botany (1788), and the arts (1785) were created. In Guatemala, the Jardín Botánico (Botanical Garden, established in 1796) and the drawing and mathematics schools (in 1797) were set up by the Sociedad Económica de Amigos del País (Economic Society of Friends of the Country). In Caracas, the Academia de Matemáticas (Mathematics Academy, founded in 1760) was supported by the business consulate. In Lima, the Laboratorio Químico-Metalúrgico (Chemicometallurgical Laboratory, 1792) was sponsored by the Tribunal de Minería (Mining Tribunal). In Buenos Aires, the Escuela de Geometría, Arquitectura y Dibujo (School of Geometry, Architecture, and Drawing) was created by the business consulate in 1799, and the Escuela Náutica (Nautical School) was created by the same organization, at Manuel Belgrano's urging. In Bogotá, the curriculum created by Francisco Moreno y Escandón for the public university (established in 1774), in spite of the university's short life, initiated interest in Enlightenment ideals and modern science. All of these institutions were established in accordance with the spirit of the century, that is, Enlightenment ideals, which focused on social reforms in science, education, and the useful arts.

In Hispanic American societies of the time, the predominant social structure was racially complex and dynamic, because the population had begun to grow at a fast clip. Because of their socioeconomic influence and their knowledge, either learned on their own or in Europe, Creoles and, to a lesser extent, mestizos took the lead in the social transformations of the eighteenth century. This allowed them to compete with Europeans for positions previously denied them in important economic activities, administration, politics, the judiciary, the church, the universities, and in cultural and scientific positions.

As the century advanced and New Spain, Peru, New Granada, Guatemala, Quito, Cuba, Buenos Aires, and so on consolidated their newly autonomous societies, there was an acceleration of cultural activity, and science was one of its expressions. Science had certainly been present throughout the colonial period, and some areas like medicine and pharmacy could be traced back to the Amerindian past. Enthusiasm for science almost took the form of isobaric lines, as pressure for it stayed constant and remarkably synchronic in American culture and involved at least three generations. The last of these generations participated directly in Hispanic American emancipation movements at the beginning of the nineteenth century. This cultural renewal was initiated and carried out by Hispanic Americans themselves or by individuals who, although born in Europe, made themselves part of Hispanic American life.

Scientific enlightenment in America, as in Europe, was a mental attitude rather that a unanimously accepted scientific or philosophical current.[8] Thus, in spite of local individual or group variants, contradictions, and eclecticism, new values (like trust in reason and experimentation as well as a search for the usefulness of knowledge) gradually took hold among cultivated Hispanic Americans, in frank opposition to values that were considered traditional or old (like authority as the source of truth, Scholasticism, and fideism). As the Enlightenment managed to permeate society—after the Creole elite carried out important informational and educational work beyond the individual sphere—society strove to transform and to adapt itself to the new political, cultural, and industrial state, which was already extant in Europe but only nascent in North America. To this end, productive sectors that followed Enlightenment ideals—such as mining—established solid alliances.

Furthermore, the material and intellectual progress desired by Hispanic American Creoles turned out to be partially in agreement with the reforming policies promoted by the enlightened despotism of the Bourbons (who had occupied the Spanish throne since 1701) and with the economic and administrative strategies adopted around 1770 for achieving a better usufruct from the colonies. Thus, there were coincidences and even common enterprises; certainly, there were also disagreements, confrontations, and polemics that emphasized the transatlantic diversity of interests and points of view regarding the region's progress.[9]

The contribution of peninsular scientists and the policies of the Spanish Crown cannot and should not be passed over, since both undoubtedly constituted influential external factors in social and cultural Hispanic American dynamics, by, for example, mounting expeditions and technical and scientific missions and by teaching. I should also mention the influence of Europeans of other nationalities, who investigated and taught in the Hispanic American territories or who became part of Hispanic American life. But we should keep in mind that this progress took place because of what Hispanic Americans themselves did. This is the subject of this chapter, and is the reason I have omitted European intervention from this discussion.

The Enlightenment can be characterized as the result of the transmigration or spread of ideas, attitudes, and knowledge that originated in France, England, and other places. Spain and its American colonies received Enlightenment ideals through Spain itself. The important thing is that the incorporation of Enlightenment ideals adopted in the Americas took the form of a graft onto a trunk with its own structures. Therefore, it is legitimate to refer to Spanish or Hispanic American Enlightenment. Although both share the traits of European Enlightenment, each possesses specificities that are comprehensible only in the light of the historical context in which they developed.

In summary, the existence of certain economic and social elements was decisive in the Hispanic American Enlightenment: powerful unions of mine owners and merchants; a dynamic domestic market; structured and diversified societies in which Creoles played the chief role; Amerindian cultural background; the population's "telluric" feeling; the need to understand an immediate reality that could not be explained by existing knowledge; Creolization and its culture. Other important aspects were nonobscurantist religiosity and Hispanic American nationalism (to be discussed later). The Hispanic American Enlightenment thus acquired its own profile, different in several aspects from that of other latitudes.

Modern Scientific Culture

In Europe between the sixteenth and the seventeenth centuries, the development of scientific knowledge accelerated and, along with its institutionalization, led to a conceptual and methodological revolution. This revolution was associated with the names of Copernicus, Galileo, Kepler, Descartes, Newton, and others, who introduced a new way to study nature. The use of instruments lent sensory knowledge an importance it had lacked, and observation and experimentation were deemed reliable procedures. The use of mathematics for expressing experimental results led to quantification and to the use of reasoning with regard to empirical data. The abandonment of theories traditionally regarded as true without any other proof than the force of Scholastic authority led to new hypotheses and concepts, which were verified with an acceptable degree of accuracy. All these innovations were synthesized in a group of rules and procedures that characterized the so-called experimental or scientific method—set forth by, among others, Sir Isaac Newton in *Mathematical Principles of Natural Philosophy* (1687) in a chapter dedicated to this topic.

Although by this time modern science was innovating and had attracted the curiosity of several investigators, in fact, it was far from acquiring the wide social recognition it would attain a century later. Scientific results were neither properly systematized nor accepted by all. There was little evidence of the advantages this revolution would bring about. Moreover, the Catholic Church had roundly condemned some of science's most famous propagandists. From a pragmatic point of view, neither results nor applications of the new knowledge were familiar, either. This forced scientists to look for alliances with other emergent social sectors and to continue their activities from the universities, schools, and other established institutions.

The marginality of modern science began to fade in Europe in the second half of the seventeenth century, when scientists received the support of the important economic, ideological, and political sectors, which were endeavoring to establish a new social order in Reformation Europe. Academies, laboratories, specialized publications, and other institutionalized forms of science began to appear. The Royal Society of London was founded in 1662, and began to publish *Philosophical Transactions;* the Académie Royale des Sciences (Royal Academy of Sciences) was founded in Paris in 1666, and began publication of *Journal de Savants* (Sages' Journal). Similar institutions were created in other countries around the same time.

In this new atmosphere, norms were imposed on scientific activity—a "decalogue," or ethos, for the scientific community. Very soon, protected by these norms, individuals like Robert Hooke, Newton, Christiaan Huygens, Edmond Halley, Gottfried Leibniz, and Nicolas Malebranche appeared, and all of them, in the favorable social atmosphere, could sum up and systematize the work of their predecessors and guide science to new accomplishments. Newton's work, in particular, provided a magnificent synthesis of the new physics as well as the pattern other sciences had to follow.[10]

At the beginning of the eighteenth century, the new science was developing in an unprecedented social atmosphere, as European governments and other sectors became interested in and supported it. As a result, new institutions corresponding to the new ideas were created in a society in transition, which was generating a new culture. Thus, scientific systematization and its spread and teaching began to be carried out inside these new institutions. Texts and magazines written in the vernacular made the new science accessible to anyone. This was one of the factors that greatly contributed to science's internationalization. The new science was at the core of the Enlightenment and was the undeniable proof of the progress it proclaimed.

In the eighteenth century, the most developed sciences were the exact ones—mathematics, experimental physics, natural history (botany, zoology, paleontology, mineralogy), geology, chemistry, and physiology—and they established the bases of the human and the social sciences. The practical benefits people expected from modern science emerged in diverse fields—medicine and pharmacy, agriculture, mining, nautical activities, geography, war, industry, and the like—which resulted in great prestige for science, its institutions, and its practitioners. A new horizon could be conceived for humankind, which meant promises of well-being and happiness for all. This was the ideology of the Enlightenment. But, as we shall see, the historical forms through which Enlightenment ideas were incorporated into Hispanic America were sui generis regarding the European model, as a result of interaction with the local social and cultural context.

Hispanic American Scientific Culture

The enthusiasm for material and intellectual progress made its way only slowly into Portuguese and Spanish America as a consequence of the isolation and restrictions imposed by the home countries. There were universities in only a few places (in Brazil, for example, there were none during the colonial period), and these obeyed the Counter-Reformation orientation prevailing in Catholic Europe as well as the most stagnant Scholasticism.[11] The printing press, the importation of books, and pedagogy in general were subjected to Inquisitorial censorship and dedicated to the transmission of established knowledge and the propagation of the faith. After the sixteenth century—which was dominated by astonishment at the nature of the recently discovered territories and by a desire to understand typical of the Renaissance and humanism (which imbued the first learned men who arrived in America)[12]—the home country was usually reluctant and opposed to modernity. This was not an obstacle for the exceptional or heterodox cultivation of science, in which scientists from New Spain and Peru stood out with their valuable mathematical, astronomical, geographical, and metallurgical studies. They were limited to the individual or small-group level, however, since their strategy for socially validating new science did not have the support of other social sectors.

How could an interest in modernity and the Enlightenment be born amid such isolation? It was a cumulative process, slow in the initial phases—because of the many difficulties to overcome—and quick at the end of the colonial period. The required impulse could only come from within the Hispanic American societies themselves, which had to mediate conflicts and negotiations between their different sectors.

In the initial phase, self-formation played an important role and was carried out thanks to the following factors: private libraries were built in spite of official restrictions and by book smuggling; publications and scientific newspapers disseminated scientific news with the double purpose of creating a culture of science and advancing scientists socially; gatherings and societies of friends of the country were formed to learn about and to transform the different Hispanic American regions. In the second phase, Hispanic American scientists therefore already had available a somewhat organized community, with a recognized ethos for their activities and some institutions they had created, but their greatest success lay in creating a leading role in society for science.

Scientific Libraries

We know that beginning in the seventeenth century there were private libraries in New Spain and Peru containing scientific works; in some cases, the lists of holdings have survived in documents pertaining to Inquisitorial activities. In Mexico, for example, the libraries of physicians Melchor Pérez de Soto and Alfonso Núñez were famous at the beginning of the century. In the second half of the century, Carlos de Sigüenza y Góngora and Sor Juana Inés de la Cruz possessed thousands of volumes, some of which were hermetic scientific texts, others, modern. Sigüenza's library contained an important collection of advanced mathematical, astrological, and astronomical texts, as well as some on physics. He also owned scientific instruments, an important collection of pre-Hispanic codices and old maps, and manuscripts in Nahuatl. These materials demonstrate the aspiration among some intellectuals to create a past for Creoles.

The "common" libraries (in schools, convents, and universities), on the other hand, held important materials, although they were basically traditional. Even the schools that belonged to the Jesuits—an order considered by some to have favored modernization—were still gathering traditional knowledge when the order was expelled from the American territories in 1767. The inventories of their libraries reveal that they lacked modern scientific texts and they were not acquainted with the works of Enlightenment philosophers.[13]

At the turn of the century, primarily from 1760 on, private libraries were the main indicator of the ever-wider circulation of books and, simultaneously, the reflection and the cause of the ideological changes taking place in the Americas. In Peru, the smuggling of forbidden books, especially from France, was so uncontrollable that in 1704 Viceroy Manuel Oms de Santa Pau was appointed and charged with allowing the French to traffic secretly in books because it was considered impossible to stop sales to Lima's residents. Censorship continued, however, and in 1708, Fray Nicolás Muñoz requested the intervention of the Inquisitor and pointed out the clandestine manner in which forbidden books were being introduced: "They have printed whole books of false doctrines, with titles belonging to Catholic authors of well-known authority. They have removed from the books of the Church Doctors what is most opposed to their perverse dogmas. They have mixed in among the works of Catholic writers great errors, which being (as they are) poison, take away the sense of the ignoramuses, and perhaps of the experts, either because of the bad inclination toward evil or because it hides under the disguise of good."[14]

In Mexico, in 1764, the Inquisition explicitly prohibited reading Voltaire and Rousseau; even those religious who had permission to read forbidden works were not allowed to do so, since Voltaire and Rousseau were heretic authors who sowed "mistakes opposed to religion, to good customs, to civil government and the righteous obedience due our legitimate sovereigns and superiors."[15] Bookstores, however, proliferated and undoubtedly were a good business, thanks to a real demand for books. In 1768, in Mexico City, there were fifteen bookstores around the city. Other businesses that combined the sale of goods and books also existed.

In fact, the spread of Enlightenment literature was associated with ideology as well as with European capitalist and industrial development. For example, the eighteenth-century's intense trade in books written in French to Hispanic America had its main seat in Cádiz, Seville, and Lisbon. Large booksellers in these cities linked with Italian, Genevan, and French publishers who distributed books in Spain and Hispanic America. An example: the accounts of Deville Brothers, booksellers from Lyon, reveal that 65 percent of their credits were in Spain and Mexico. Even some French booksellers—or of French origin—moved to Hispanic America to develop their business there. A certain Agustín Dherbe, a book merchant in Mexico City, announced, in a thick catalog, 4,336 titles for sale.[16]

There were, of course, other booksellers moved mainly by ideological concerns. Such was the case of Father Diego Cisneros, who, shielded by the privileges his position offered him, opened a bookstore in Lima from which, besides selling missals and other religious books, he sold books by modern philosophers and by the Encyclopedists. Fray Diego Cisneros was, certainly, a strong supporter and defender of the Peruvian Enlightenment and, in 1794, he used his own money to pay for the publication of volume 12 of *Mercurio Peruano*, the first scientific magazine in that country.[17]

The private libraries from the 1760s whose contents have survived are a clear expression of the concerns of Hispanic Americans who supported Enlightenment ideals, of the era's ideological debates, and of the scientific, technical, and cultural changes in the Americas. Let us look at some examples. New Spain physician and mathematician José Ignacio Bartolache, for example, left at his death a library consisting of 487 works published in 712 volumes. He owned books written in Latin, Greek, Hebrew, Nahuatl, English, and French (21 of the last). His collection included 80 literature books, 75 on medicine, 60 on religion, 50 on law, 25 on mining, 21 on chemistry, 20 on history, 20 on physics, 15 on mathematics, 16 on botany and the natural sciences; the remainder were dictionaries or about geography, travel, music, philosophy, indigenous and European languages. His library boasted 177 books on science, including *Mathematics*, by Christian Wolf, in five volumes; *Elements of Chemistry*, by the Académie de Dijon; *Essay on Metallurgy*, by Francisco Sarria; *Theory of Light*, by Antonio Lequio; *Newtonian Physics*, by Voltaire; and *On the Structure of the Human Body*, by Vesalius. The contents of his library allow us to observe that, for Creoles who espoused Enlightenment ideals, modern scientific culture—in addition to traditional culture (i.e., religion, law, literature, and so on)—occupied an important place in their intellectual formation. If we add to this the scientific instruments Bartolache possessed (among them, microscope, hydrometer, magnifying glass, and thermometer), it is clear that his interest in the new science was both theoretical and practical.[18]

Bartolache's case is not exceptional either in Mexico City or in other Hispanic American capitals. The inventory of New Spain mathematician and astronomer Antonio de León y Gama shows he possessed a library of some seven hundred works. The scientific topics include mineralogy, medicine, chemistry, astronomy, geography, physics, engineering, botany, mathematics, architecture, numismatics, and topography. He was also apparently interested in pre-Hispanic antiquities and indigenous chronicles and languages.

In Peru, important private libraries were owned by Pedro de Peralta y Barnuevo, an engineer and astronomer; José Dávalos, a mulatto doctor who studied in Montpellier, France; Hipólito Unánue, doctor and naturalist; Toribio Rodríguez Mendoza, ecclesiastic and educational reformer; José Baquíjano y Carrillo, a lawyer and scholar who built his collection in Europe; and the above-mentioned Cisneros. It was possible to find in these libraries scientific texts and works by Enlightenment philosophers, and, as was the case with private libraries in New Spain, they reflected a lively interest in local pre-Hispanic culture.

In Peru, this interest in indigenous matters not only involved historical scholarship but also was practical. It led Diego de Avendaño to publish in 1668 the *Thesaurus Indicus*, a work that became fundamental for everything referring to the Indies and that advocated for Indians and blacks. Baquíjano's case is remarkable because, in 1781, before Viceroy Agustín de Jáuregui, he bravely and in public censured the cruelty of the Spaniards in putting down the indigenous uprisings of Huanuco, Arequipa, and Urubamba after defeating Tupac Amaru, who was drawn and quartered in Cuzco on *visitador* José Areche's orders.

Equally important for their modern content (as well as for their natural history collections) were the libraries owned by physicians and naturalists Eugenio Espejo and José Mejía Lequerica in Quito; physician Tomás Romay in Cuba; Dean Gregorio Funes in Córdoba; publisher and writer José Antonio Alzate in New Spain; and doctor and naturalist José Celestino Mutis and naturalist and astronomer Francisco José de Caldas in New Granada. We know the content of Caldas's library because of the inventory made when it was confiscated in 1816. It consisted of 94 works in 164 volumes, besides "many papers and amassed maps and two cases of mathematics." Among other scientific books were *Chimie*, by Baumé; *Astronomy*, by Jérôme Lalande; *On Mathematics*, by Wolfio; *Encyclopedia*, first volume; *Nova Genera Plantorum*; *Treaty of Cosmography*; as well as several manuscripts recording his astronomical and botanical observations in Quito and Santa Fe de Bogotá. Interestingly, besides scientific books, Caldas's library, like most other libraries belonging to scientists at the end of the eighteenth century, contained works on technical and applied science, for example, treatises on agriculture, military architecture, artificial canals, grains, astronomy, navigation, and artillery exercises.[19]

This interest in the "useful arts" is characteristic of Hispanic Americans who adopted Enlightenment ideals in the last third of the eighteenth century, when the Enlightenment had already been assimilated and people were trying to use it to learn about and transform their own country. In Mexico in 1795, two cases containing eighteen books—five about social topics, two on chemistry, one bill of exchange, and ten on technique—arrived for the bookseller Mariano de Zúñiga y Ontiveros. The books on technique concerned machinery, mulberry cultivation, wax craft, calico dyeing, brass production, canvas whitening, paper production, silk dyeing, hat making, and barbering.[20]

Thus, to a basically individual and scholarly interest in science was added Hispanic Americans' concern with carrying out a social reformation based on science and the useful arts. At this moment in time, scientists began to play a social role and found other sectors with which to talk: artisans, mine owners, merchants, viceregal bureaucrats, and so on, with whom they negotiated strategies for making viable their cognitive and practical purposes.

Scientific and Technical Journalism

The formation of a scientific culture was assisted by efforts to popularize science. Although pamphlets, handbooks, and books were also used for this purpose, Hispanic Americans who believed in the Enlightenment, individually and collectively, primarily utilized scientific and technical journals.

The first properly scientific magazine of the American Enlightenment, *Diario Literario de México* (Literary Magazine of Mexico, established in 1768), was published by a citizen of New Spain, José Antonio Alzate y Ramírez. This intellectually curious scientist and Creole writer took on the enormous task of popularizing science over the next thirty years by publishing, besides the *Diario Literario*, *Asuntos Varios sobre Ciencias y Artes* (Miscellaneous Science and Art Subjects, 1772–1773), *Observaciones sobre la Física, Historia Natural y Artes Útiles* (Observations on Physics, Natural History, and the Useful Arts, 1787–1788), and *Gacetas de Literatura de México* (Mexican Literary Gazettes, 1788–1795). Alzate's scientific and popularizing works had great repercussions in Mexico as well as in other places in the Americas and Europe (he was, e.g., elected corresponding member of the Académie Royale des Sciences of Paris).

Because the *Diario Literario* was the first science journal, it is worth reviewing its goals and content. It published only eight numbers over the course of three months in 1768, because viceregal orders, alleging "just cause," suppressed it. According to Alzate in the first issue, the newspaper would imitate the three customary areas of European newspapers: it would review all types of literary works; it would expound on physics and mathematics; and it would discuss "economic" works on "agriculture, trade, navigation, and everything related to the public welfare." Alzate announced that special attention would be paid to local topics such as agriculture (because "it needs much improvement"), mining (subjected to "blind practice"), American geography ("so unknown"), natural history ("because of its particularity and because the authors that have written about America have never mentioned it"), and medicine. Finally, he invited his readers to critique him, to make suggestions and comments, and to send him news to publish.

Excerpts of works by other authors were published in issues 2 and 3 of the *Diario Literario*, with notes written by Alzate; these notes indicate his erudition, since he quotes multiple modern scientists. These texts treat theological subjects and express the peculiar problems of the Hispanic American Enlightenment, that is, the reconciliation of modern science and religion. The fourth issue contains a geographical description of the

Province of Sonora prepared by Alzate and an astrological chart. The fifth issue is about the steam engine (the first detailed information in Hispanic America about this piece of machinery) and the advantages of using it for draining mines. The sixth issue describes an earthquake felt in Mexico City and the different classes of earthquakes, their origin, and so on. The seventh issue contains texts on the cultivation of cacao and its processing and the "convenience of pocket watches." Finally, in the eighth issue, a letter was published on theater, its reformation in Europe, and its utility in New Spain.

New publications of this type followed the demise of *Diario Literario de México*. First to appear was *Mercurio Volante, con Noticias Importantes y Curiosas sobre Física y Medicina* (Flying Mercury, with Important and Curious News about Physics and Medicine), published by José Ignacio Bartolache. This weekly newspaper was published in Mexico from October 17, 1772, to February 10, 1773, for a total of sixteen issues. This was the first magazine dedicated to medical topics, and it was established almost simultaneously with Alzate's second magazine, *Asuntos Varios sobre Ciencias y Artes*, which ceased publication after thirteen issues. Bartolache, like Alzate, intended to write for the masses and not for specialists and used Spanish for that reason. Both men hoped this would help overcome the cultural backwardness of New Spain.[21]

From the start, Bartolache expressed his interest in the new science, and in the second issue he praised experimental physics and Newton. He wrote on the following topics: the importance of the thermometer and the barometer; the medical arts; hysteria from the medical and psychological points of view; the history of pulque (a pre-Hispanic drink) and its physical and chemical composition; and the importance of anatomy in medicine.

Other magazines followed. *Advertencias y Reflexiones Varias Conducentes al Buen Uso de los Relojes Grandes y Pequeños y su Regulación: Papeles Periódicos* (Miscellaneous Warnings and Reflections on the Proper Use of Large and Small Clocks and Their Regulation: Occasional Papers) was also published in Mexico, by Diego de Guadalajara beginning in 1777. It was dedicated to chronometry and the construction of instruments. *Observaciones sobre la Física, Historia Natural y Artes Útiles* was published, as mentioned earlier, by Alzate along with *Las Gacetas de Literatura*, which began publication while *Observaciones* was still active.

Starting with these precedent-setting periodicals, public scientific and technical journalism quickly developed in the major cities of the continent simultaneously with improvements in how it was popularized and what it covered. Even established gazettes and other general periodicals began to incorporate scientific and technical news and writing. Furthermore, associations created by proponents of Enlightenment ideals (with the participation of other sectors) began to support the individual efforts of pioneers to maintain scientific publications. Thus, for example, around 1787, a group of scholars concerned with the cultivation of "enlightenment" in their homeland met at the Academia Filarmónica (Philharmonic Academy) in Lima; they later created the Sociedad de Amantes del País (Society of Lovers of the Country). The members of this group, besides holding meetings and discussions, decided to publish a magazine entitled *Mercurio Peruano*, whose editor was Jacinto Calero. The first issue appeared on January 2, 1791. A total of twelve issues were published biweekly and addressed such topics as botany, agriculture and cattle, medicine, mineralogy, physics, Peruvian history and ethnography, social and economic matters, teaching reforms, and theology. *Mercurio Peruano* also contributed to the popularization of Enlightenment philosophers such as Voltaire and Rousseau (although sometimes the paper criticized them) and scientists such as Johannes Kepler, Newton, Leibniz, Wolf, John Locke, Bernard de Fontenelle, and Pierre Bayle. Neither political motivation nor revolutionary propaganda, however, can be found in the paper's pages. As in other Enlightenment newspapers, what mainly appears is Creole opposition to the official peninsular culture in the form of articles on their own culture.

Brilliant Creole intellectuals contributed to this magazine, for example, Unánue, Baquíjano, Cisneros, Calero, Rodríguez, Rossi, Cerdán, and Calatayud. Among the most significant articles were those signed by "Cefalio" (Baquíjano) and "Aristio" (Unánue). The following articles, written by Unánue, contain valuable data and scientific observations: "Dissertation on the Cultivation, Trade, and Virtues of Coca," "Scientific Description of the Plants of Peru," and "Observations on Lima's Climate." Unánue's works gained the recognition of his contemporaries and of the scientific societies of, among other places, Philadelphia, New York, and Bavaria, all of which named him a corresponding member.

The following scientific and technical magazines were published before the end of the eighteenth century because of the initiative and support of economic societies of friends of the country: *Primicias de la Cultura de Quito* (First Fruits of the Culture of Quito, 1791); *Memorias de la Sociedad Económica* (Havana) (Memoirs of the Economic Society, 1793); *La Gaceta de Guatemala* (The Guatemala Gazette), which started up again in 1797 with the help of the Sociedad Económica de Amigos del País (Economic Society of Friends of the Country); and in 1801, the *Telégrafo Mercantil, Rural, Político-Económico e Historiográfico del Río de la Plata* (Mercantile, Rural, Politico-Economic, and Historiographic Telegraph of Rio de la Plata). These publications were encouraged by the Enlightenment philosophy that characterized the economic societies: to study the country; to promote reforms in education and major economic activities; and to modernize scientific and technical areas.

As part of the nineteenth century, but also encouraged by Enlightenment principles, Creole Francisco José de Caldas began publishing the important *Semanario del Nuevo Reino de Granada* (New Kingdom of Granada Weekly) on January 3, 1808. It appeared in weekly broadsides in 1808 and 1809, and later in monthly signatures or memoirs on particular topics—of which eleven were printed. Among other subjects, the *Semanario* published works on agriculture (corn, wheat, nutmeg, potatoes, cacao, etc.), industry, statistics, roads, navigable rivers, mountains, soil crops, the exact sciences, eloquence, and history. According to Caldas, the newspaper was of general interest: "Bishops, governors will find much enlightenment for successful rulings; the economist, the farmer, the geographer, the merchant will pick up knowledge that either does not exist today or is in the manuscripts of men of letters and would not see the light of day if the *Semanario* did not exist." For this reason, he called on "men of letters and good patriots" to support the publication by subscribing to it and by writing for it. The response was rapid, and New Granadan authors, among them, Eloy Valenzuela, José de Restrepo, José Manuel Campos, and José Joaquín Camacho, published important works in the paper.

The following scientific and "economic" magazines were published in Mexico at the beginning of the nineteenth century: *Semanario Económico de Noticias Curiosas y Eruditas sobre Agricultura y Demás Artes, Oficios, Etcétera* (Economic Weekly of Curious and Learned News about Agriculture and Other Arts, Occupations, Etcetera), established by Wenceslao Barquera in 1808, 108 issues; and *El Mentor Mexicano: Papel Periódico Semanario sobre la Ilustración Popular en las Ciencias Económicas, Literatura y Arte* (The Mexican Mentor: Weekly Publication on Popular Enlightenment in the Economic Sciences, Literature, and Art), 1811, 48 issues. As with the other cases mentioned, this kind of journalism shows clearly the new social role science was playing in Mexico—very different from the, for all intents and purposes, isolated and "Quixotic" position prevailing three decades earlier.

I have already mentioned that this interest in the popularization of science and scientific publication expanded to nonspecialized publications. The *Papel Periódico de La Habana* (1790) published scientific news and articles written by authors from Cuba and other American countries; it also expressed the opinions prevailing in the Sociedad Económica de Amigos del País. The same thing happened with other Latin American magazines, such as the *Semanario de Agricultura, Industria y Comercio* (Agriculture, Industry and Trade Weekly, 1802), edited by Hipólito Vieytes in Buenos Aires; the *Gaceta de Caracas* (Caracas Gazette, 1808), published by Jaime Lamb and Mateo Gallagher after the introduction of the printing press in Venezuela; *O Patriota* (The Patriot, 1813–1814), published in Rio de Janeiro by Manuel Ferreira de Araújo; *El Correo de Comercio* (The Trade Mail, 1810), founded by Manuel Belgrano in Buenos Aires; and other publications that regularly appeared in, for example, Mexico City, Puebla, Veracruz, Lima, Arequipa, Quito, Santa Fe, Santiago, and San Juan.

The evolution of scientific literature between 1768 and 1810 allows us to track the course of the lively ideological debate carried out by Hispanic Americans who accepted Enlightenment ideals against Scholasticism and traditional knowledge. At the same time, the gradual introduction of modern scientific thought (from the likes of Copernicus, Newton, Georges Buffon, Carolus Linnaeus, and Antoine Lavoisier) can be observed, along with the intense polemics between Creole scientists (e.g., Alzate, Unánue, Bartolache, Espejo, Mejia, and Caldas) and Spaniards and Europeans (e.g., Manuel Martí, Cervantes, De Pauw, Reynal, and Robertson) in an attempt to defend American scientific culture, history, and nature in the face of repeated scorn, attacks, and slander.[22]

Hispanic American scientific magazines also helped broaden the Creole Enlightenment's influence on sectors of the population involved in the task of reform. As a result, in the areas of education, culture, agriculture, mining, and industry, diverse reforms were introduced, for example, the gradual abandonment of Scholasticism in teaching; the rescue and dissemination of languages and other aspects of native culture; measures to improve indigo, mulberry, cotton, and tobacco crops, among others; and innovations in mining and other industrial areas.

By using science and the useful arts, Creoles who supported Enlightenment ideals proposed or introduced the reforms they deemed appropriate for the reality they knew directly. In so doing, several times they had to oppose the Spanish government's authoritarian initiatives and show that their viewpoints were correct or even superior. In Mexico, for instance, Alzate presented suggestions in his magazines for improving the extraction of minerals and water from mines, for ventilating mines, for ginning cotton, for cultivating and processing cochineal. In Peru, Unánue published statistical data on public health and his botanical studies of Peruvian plants. In New Granada, Caldas wrote about how to cultivate cochineal, the adaptation of the vicuña, the relationship between the geography of Santa Fe and economy and trade. In Buenos Aires, Vieytes proposed reforms to the agriculture of Río de la Plata and the use of chemistry in agriculture and industry. In Guatemala, a Mexican botanist, Mariano Mociño, proposed measures for improving the cultivation of indigo, and Tomás Zelaya and others presented suggestions regarding linen and cotton. Such examples abound and allow us to appreciate one of the essential characteristics of Hispanic American Enlightenment science, namely, its pragmatic orientation.

Scientists sometimes had to use their knowledge for resisting the measures the Spanish government tried to impose beginning in 1770. Indeed, Bourbon reforms were meant to increase the economic exploitation of the Hispanic American colonies and subject them to strong administrative, fiscal, political, and military control. Hispanic American opposition to these reforms, especially that of Creoles and mestizos, whose interests were most affected, was remarkable. In Peru and Mexico, Creoles and mestizos refused to introduce a new method for processing silver (named after its European "inventor," Baron Born, although, in fact, it was known in the Americas as early as the seventeenth century).[23] This project was encouraged by distinguished, and arrogant, mineralogists sent by the home country: Fausto de Elhúyar, from Spain to Mexico; and the Baron von Nordenflicht, from Sweden to Peru. Both men had the support of a group of competent German technicians. After attempts to demonstrate the efficacy of the barrels method or the Born method, the superiority of the patio process, discovered by Bartolomé de Medina in Pachuca (Mexico) in the sixteenth century, was proven. After ten years, the Europeans recognized that the processing of minerals containing little silver or gold was more efficiently done by means of amalgamation (the patio process), the method Americans traditionally used. Before the foreign mineralogists began work, in 1787, José Antonio Alzate of New Spain published in *Observaciones* a solid argument in favor of the patio process and pointed out that the Born method had been known by Americans since the seventeenth century, after Álvaro Alonso Barba wrote about it in his *Arte de los metales* (Art of Metals, 1640).

Enlightenment scientific newspapers rendered possible communication among scientists in each country and also—and this is very important—among those from different parts of the Americas. This is evident in the correspondence between readers and publishers of these magazines, in the variety of authors, in the ongoing debates, and so on. Intercountry communication, revealed in citations to and the reproduction of articles published in other Hispanic American magazines, shows a solidarity of ideals and the gradual formation of the Hispanic American "Republic of Science." Several works by Alzate were published in Lima and Santa Fe, and articles from the *Mercurio Peruano* were reproduced in Havana, for example.

These publications also established scientific approaches, practices, values, and methods, all of them at variance with the frozen practices of Scholasticism. At the same time, a scientific ethos and a normalization of activities were established, prior to the institutionalization of science. The scientific community, poorly covered in the press until then, could use periodicals as an appropriate instrument for popularizing its works, for learning about what was going on in the Americas, Europe, and in the nascent North America (the last of which provoked great interest regarding scientific matters, as seen in Alzate's praise for Benjamin Franklin's work).

The almost constant appearance of articles about Hispanic American geography, natural resources, culture, economy, and history in the pages of Enlightenment periodicals, as well as about the possibilities of autonomous development, contributed to the formation of a national conscience in the Hispanic American nations. Scientific nationalism was added to Creoles' telluric and patriotic feeling. Both were joined to produce a clearer awareness of the geocultural reality they faced daily. In the end, Hispanic Americans' gradual process of discovery of their own historical identity inevitably led them to emancipation from Spain.[24]

Modern Scientific Paradigms in Hispanic America

As I have pointed out, it was during this period that modern scientific paradigms began to spread and were assimilated by colonial society. This happened through self-study; education, journalism, and institutions dedicated to this purpose. The spread of modern scientific theories, particularly those dealing with physics, astronomy, and mathematics, had a remarkable history in the seventeenth-century Americas. Their assimilation began belatedly, however, in the middle of the eighteenth century and picked up steam in the last third of that century. From that moment on, a remarkable updating of knowledge took place, and interest in practical applications and research grew, in some areas (e.g., chemistry, metallurgy, and mineralogy), contemporaneous with European research. Linnaean taxonomic systems, modern botany in general, other branches of natural history, as well as Newtonian physics spread and were cultivated intensely beginning in the 1770s and the 1780s.

Simultaneously, the "domestication" of science by integrating it into and adapting it to the Hispanic American context took place. Science became more pragmatic, and scientific knowledge became more localized in terms of education and ideology, as we shall see.[25] In the beginning, isolated individuals popularized modern science, and they frequently had to act outside of established institutions (universities and religious schools) and even secretly; later, science began to be taught and promoted by innovative (i.e., lay) institutions.

As was to be expected, the main Hispanic American mining centers had the benefit of dynamic socioeconomic conditions that led to modernity early on. Technical matters were of interest in these regions, particularly if they related to excavation and drilling, tunnel construction, drainage, metals processing, and the minting of money. There were, however, other kinds of discoveries, such as the hypsometer, created by Caldas; the platinum ore in the mines of Chocó; erythronium (vanadium), discovered by Del Río; the chemical composition of the upper atmosphere, detailed in the celebrated *Gacetas de Literatura* by Rangel and Alzate in their explanation of boreal aureoles.

Modern science was present around Charcas from 1630, where Álvaro Alonso Barba, author of the important and celebrated *Arte de los metales*, accepted as true Galileo's discoveries of "seeing instruments or for seeing from afar." This is remarkable because of Galileo's condemnation, which should have been a warning to potential followers. Later, Nicolás de Olea, from Lima, was the first man in Peru to question official Thomism and made the first references to Renaissance and modern authors such as Campanella, Bruno, and Brahe. He also spread Cartesianism, whose natural philosophy was easier to reconcile with the prevalent Aristotelian-Thomist tradition. At the end of the seventeenth century, at the Universidad de Chiquisaca, José de Aguilar, of Lima, wrote *Curso de filosofía* (Philosophy Course, published in 1701), in which he declares himself prone to Cartesian principles and daringly asserts that stars move because of extrinsic forces and not "angelic impulse."

In the first half of the eighteenth century, the most important personality in South American science was José Eusebio Llano Zapata, also from Lima. Among other works, he published "Naturaleza y origen de los cometas" (Nature and Origin of Comets), "Verdadero modo de conservar la salud" (True Way of Preserving Health), "Observación diaria-crítica-histórica-meteorológica" (Daily-Critical-Historical-Meteorological Observation), and "Memorias histórico-físicas-críticas-apologéticas de la América meridional" (Historical-Physical-Critical-Apologetic Memoirs of South America). His work shows the author's interest in the meticulous study of nature and presents the independent criteria he used in his research. Llano Zapata was also responsible for the revolutionary initiative for creating a school of metallurgy in Peru, as well as a public library to stimulate the popularization of science. The opposition of a society resistant to change, however, caused the miscarriage of these projects. Llano Zapata's criticism—in letters to friends—of the social and political vices of the viceroyalty are most interesting for understanding what was happening in colonial Peru by that time.[26]

In Mexico, at the other end of Spanish America, in the third decade of the eighteenth century, the interest in modern science began in a small, organized discussion group. This group certainly showed a desire for knowledge but also the early Creole and mestizo intellectual opposition to and uneasiness with the Spanish commitment to orthodoxy.[27] In 1648, several Inquisitorial actions began against some of the members of this group, for example, Guillén de Lampart, because of his ideas on independence and his scientific heterodoxy, and Melchor Pérez de Arboleda, for possessing forbidden books and practicing astrology. The persecution this seventeenth-century community suffered—their books were confiscated; they were censored and subjected to Inquisitorial trials—explains why the scientific nationalism expressed so strongly one century later was born here.

Fray Diego Rodríguez of New Spain, a Mercedarian, was the visible and respected head of this group. He held the astrology and mathematics chair at the Facultad de Medicina (Medical School) of the Universidad de México in 1637, and from there he disseminated the mathematical and astronomical theories of Copernicus, Brahe, Galileo, Gilbert, Tartaglia, Besson, Cardano, Neper, and others. He wrote several treatises on these matters, but the most important are his studies on third- and fourth-degree equations, logarithms, and the application of these to astronomical calculations as well as his studies on spherical trigonometry and chronometry. He carried out several engineering works and questioned Aristotle's authority, too. In *Discurso etheologico del nuevo cometa* (Ethereological Discourse on the New Comet, 1652) he supports the celestial origin of comets. He exercised intellectual influence for a long time and in different places (e.g., Peru, the Philippines) through his students and his work.[28]

Carlos de Sigüenza y Góngora, also from New Spain, held the mathematics chair at Mexico's university and espoused modern scientific methods in mathematics, astronomy, cartography, and physics. He wrote important works on engineering, agronomy, and Indian chronology, and he engaged in a famous debate with the German Jesuit Eusebio Kino on the supposedly malicious influence of comets—on account of the appearance of one in 1680. In this debate, he demonstrated his modern ways of thinking and astronomical knowledge as well as precise astronomical calculations (carried out at the same time as Newton's).[29]

There were others in New Spain who concerned themselves with science throughout the colonial period. These forward thinkers influenced people in other parts of the Americas, for example, Marcos Antonio Riaño Gamboa, a Cuban by birth, who studied medicine in Mexico and temporarily succeeded Sigüenza as mathematics chair at the university. Returning to Cuba, he participated in the founding of the Universidad de la Habana and carried out the first systematic astronomical observations applied to geography on the island. His results were published in the memoirs of the Académie Royale des Sciences of Paris and praised by Cassini in 1729 for their precision and by the students of Diego Rodríguez.[30]

It is worth noting that tradition was present in these early Hispanic American modern scientists. José de Aguilar, Diego Rodríguez, Sigüenza y Góngora, and Llano Zapata, for example, tried to reconcile their modernity with other beliefs firmly established in their environment at the time and therefore difficult to avoid. Aguilar sustained Ptolemaic points of view in astronomy, and Rodríguez, although a supporter of heliocentrism, veiled his beliefs because of the threats to heterodox scientists. Sigüenza, for his part, held hermetic beliefs. It should be remembered that in that century the coexistence of different scientific paradigms even in one individual was common not only in Hispanic America but in Europe, too.

The attempt to reconcile not only different scientific paradigms but also science and religion is a characteristic of the American Enlightenment. Indeed, in the Americas, atheism and materialism—characteristics of the French Enlightenment—were rare. There is also little evidence of the spread in Hispanic America of the scientific materialism of Diderot, La Mettrie, or D'Holbach (although Juan Antonio de Olavarrieta is well known in Mexico for writing "El hombre y el bruto" [The Man and the Beast] at the end of the eighteenth century, inspired by *Système de la nature*, by D'Holbach).[31]

Furthermore, there were numerous metaphysical treatises, philosophy courses, and physics texts written by Hispanic Americans trying to reconcile theology, metaphysics, and science by shielding themselves in eclectic positions or resorting to probabilism. The latter affirmed that "it is reasonable to follow the truly probable opinion; the less-probable opinion, compared with the most probable, is probable, truly: therefore, it is reasonable to follow it." Under this reasoning, Copernican or Newtonian theses were sustained as *probable;* with regard to social matters, regicide and tyrannicide were regarded as probable by probabilistic thinkers. It was the Jesuits, mainly, who came out in favor of probabilism and against sectarianism, the latter of which prevented them from developing new forms of intellectual hegemony (based on modern science and philosophy) for the Jesuit Order. Other orders, like the Dominicans, the Franciscans, and the Augustinians, opposed probabilism and succeeded in having it outlawed in Peru around 1762 and in other parts of the Americas about the same time. Finally, the Jesuits were expelled from the American territories in 1767.

It is interesting to observe the effect of nonobscurant religion on Hispanic Americans who adopted Enlightenment ideals when religion combined with their scientific convictions. Nonobscurant religion was a mediating and helpful element for adopting daring and radical viewpoints, such as the ones modern science and the Enlightenment represented in Hispanic American society. This was, for example, the effect attached to devotion to the Guadalupana (the "brown Virgin") by Creoles in New Spain who believed in Enlightenment ideals. In Mexico, devotion to the Guadalupana constituted a truly nationalist ideology with broad social acceptance in the religious and cultural spheres. That is why scientists such as Bartolache and Alzate concerned themselves with this topic. Besides protecting them against charges of nonreligiosity or heresy because of their commitment to modernity, their study of the Guadalupana helped them show that religion and science were not incompatible. Their project of scientific nationalism could thereby make its way through the superstitious and religious mentality of the time.[32]

The initial stimulus the pure and applied sciences received in the Americas during the seventeenth century continued uninterrupted in the following one. Moreover, as the eighteenth century advanced, other regions gradually adopted the impetus. Certainly, economic and social aspects played a decisive role, as mentioned at the beginning of this chapter (inasmuch as they continued to be developed), in the progress that began in the second half of the century; we should therefore keep them in mind.

In the Río de la Plata region, Buenaventura Suárez, of Santa Fe, a graduate of the Universidad de Córdoba, installed an astronomical observatory in 1706, equipped with telescopes and precision clocks, some of which he built himself. His work on the periods of the satellites of Jupiter was known and appreciated in Europe. He also wrote *Lunario de un siglo* (A Century's Moon Recordings) around 1744. Suárez's case is rare and for that reason his interest in observation and the use of instruments is remarkable as an example of autodidacticism. In the Viceroyalty of Río de la Plata, scientific modernization was delayed until social conditions were favorable, that is, until independence (1810).

Jesuit José Gumilla, a Spaniard who studied in Santa Fe de Bogotá and who wrote in Venezuela *El Orinoco ilustrado y defendido* (The Orinoco Illustrated and Defended, 1741), provides a similar case. In this work, the natural and civil history and the geography of the Orinoco River region are described in modern overtones—the author spent twenty-five years in his order's missions in that area.

In New Granada, the teaching of mathematics and modern physics started in 1762, as a result of Creole initiatives aimed at forming an elite that possessed knowledge useful to the functioning of the state. These initiatives made good use of José Celestino Mutis, the viceroy's doctor, who taught the first modern mathematics and physics courses at the Colegio de Nuestra Señora del Rosario. Later, he was replaced by his New Granadan students Fernando Vergara, Jorge Tadeo Lozano, and Francisco José de Caldas (although others held the position between his leaving and his students' being appointed). In 1773, Mutis also taught the first-ever courses on Copernican astronomy in New Granada, which led to a quarrel with the Dominicans of the Universidad de Santo Tomás, who attacked him for contradicting the astronomy doctrine authorized by the Inquisition.[33] In order to start teaching Newtonian physics, Mutis began the first known translation into Spanish of a part of Newton's *Principia* (1770), as he had some recourse to a slanted version of this work, one that allowed him to adapt the new theoretical paradigm of physics to the limited institutional, ideological, and social conditions of New Granada.[34]

Mutis also introduced Linnaean botany in 1782 and reforms in the teaching of medicine in 1805. The former resulted from the Expedición Botánica of the New Kingdom of Granada, which he was responsible for starting, with the collaboration of Creoles Eloy Valenzuela and Antonio García. The latter were instituted when the professorship in medicine was reestablished in the Colegio de Nuestra Señora del Rosario to present a modern viewpoint that combined clinical medical teaching with auxiliary sciences (mathematics, physics, chemistry, and botany).

The work of Mutis in New Granada was most important, and it lasted nearly forty years, from his arrival in the Americas until his death in Bogotá in 1808. He was truly the initiator of the scientific movement in that region and the mentor of a generation of eminent Creole scientists.

His assimilation into the country that received him was complete. Mutis organized the Sociedad Patriótica in 1801, and he always fostered an attachment to the country in his numerous followers. In 1791, Francisco Antonio Zea, a pupil and collaborator, echoing New Granadan youths who supported Enlightenment ideals, wrote in the *Papel Periódico de la Ciudad de Santa Fe de Bogotá* (Weekly Paper of the City of Santa Fe de Bogotá) under the title "Avisos de Hebephilo" (Warnings of Hebephilo), a harangue to "the youths of the two schools [in the capital] about the uselessness of their current studies [, the] necessity to reform them [, and the] selection and good taste in those they should embrace." Like his fellow students who espoused Enlightenment ideals, Zea aspired to be a "sage of the republic" in order to use science, by institutionalizing it, for the economic exploitation of the country and as a source of political power for the Creoles. Some other students (Zea, Nariño, Caldas), facing the negative response of authorities to their proposals for reform, became radical years later and headed up the revolutionary road.[35]

As in the rest of the colonies, well into the eighteenth century, official teaching institutions in Peru, like the Universidad de San Marcos, continued within the strictest intellectual orthodoxy and lagged behind in scientific advancement. Even the chairs of mathematics and medicine, which might have developed at the university, had disappeared. Their reintroduction was extremely difficult, and when it was finally possible, it was because of the zeal of Peruvians who had adopted Enlightenment ideals at the end of the century. The absence, furthermore, of academic institutions besides universities hindered the spread of modernity. Finally, such spread was the result of actions taken by Creoles from outside who took up Enlightenment ideas.

To correct this situation, José Baquíjano and a group of like-minded Peruvians struggled to renovate university studies in Lima.[36] After their initial failure to influence the frozen university bureaucracy, this group proposed to take their reform program to the Colegio de San Carlos or the Convictorio Carolino. Although the Colegio was attached to the university, it had nevertheless acquired a certain autonomy, which represented a refuge and freedom to teach for Enlightenment Hispanic Americans.

In 1786, a Creole from Chapapoya, Toribio Rodríguez de Mendoza, was appointed interim president of the school. This marked a new age for this institution, since now it could spread the Enlightenment, within limits, to the intellectual life of Lima. It started teaching the Newtonian system in Peru, although in an incoherent environment in that the study of the moderns was allowed but Aristotle's texts had to be used in public and in competitive exams. This is why the rector declared, "And who does not see that it is a highly ridiculous monstrosity to force the Caroline [students], who profess to study the new philosophy, to defend Aristotelian doctrine?"[37]

In Lima, another very important step in reforming and teaching science took place in 1791, when Creole doctor Hipólito Unánue established the Colegio de Medicina y Cirugía (School of Medicine and Surgery). Besides the modern education of doctors, this school dealt with hygiene and public-health problems.[38]

In Cuba, the scientific reform also was begun by the action of men who had adopted Enlightenment ideals. Two of them played an important role: cleric José Agustín Caballero, who unsuccessfully attempted to reform the university and to introduce the teaching of modern philosophy; and physician Tomás Romay, an enthusiastic promoter of scientific medicine, the natural sciences, and chemistry. Both were linked to the nascent bourgeoisie, which expressed itself in the Sociedad Patriótica and the *Papel Periódico de La Habana*, the two instruments of Cuban modernity since 1792. In 1814, Félix Varela began a systematic popularization of physics and chemistry through his writings and teaching.[39]

I have already mentioned the role the Sociedad Económica played in Guatemala in reform projects. People associated with it effected change with their scientific works. Botanists José Mariano Mociño and José Longinos Martínez arrived in Guatemala from New Spain and promoted the natural sciences and the establishment of a botanical garden. The local or resident Creoles who stood out were Jacobo Villa Urrutia, who was the first president of the Sociedad Económica; José Antonio de Liendo y Goycochea, who carried out reforms in the university and introduced the study of modern physics and the experimental method; a physician from Chiapas, José Felipe Flores, royal physician, professor, and reformer of the teaching of medicine, besides being a botanist and experimental physicist; Narciso Esparragosa, a student of Flores's and a brilliant surgeon. The list goes on.

At the Real Audiencia of Quito, the popularization of modern scientific theories was also associated with the actions of Quito's residents themselves. The first person to spread Copernican, Cartesian, and Newtonian theories was Quito Jesuit Juan de Hospital, between 1759 and 1762. Some of his students—Eugenio Espejo (physician), Miguel Antonio Rodríguez (philosopher), and José Mejía Lequerica (botanist)—continued this work with clearly ideological purposes besides the pragmatic ones already described. They used science as an instrument for developing self-awareness about the historical space and time of Quito and to get a first glimpse of the national reality. These men continued the earlier work of Pedro Vicente Maldonado and Juan de Velasco regarding geography (the former) and the history and ethnography of the country (the latter).

The Brazilian case differs considerably from what happened in the parts of the Americas controlled by Spain. As has already been mentioned, Brazil did not have universities, and intellectual life in the colony was minimal. This led the Creole elite to go to Portugal and other European countries to study. King José I of Portugal, influenced by the European Enlightenment, authorized his minister, the Marquis of Pombal, to introduce important reforms in the kingdom, particularly in the Universidade de Coimbra in 1772. Several Brazilians studied science in Portugal and engaged in important scientific activities in Brazil, Bartolomeu de Gusmão, the builder of aerostats, for example. Others occupied important public positions in Portugal and in the Portuguese African colonies. Such was the case of José Bonifacio de Andrada e Silva, a remarkable mineralogist who spent thirty years in Europe in contact with numerous scientists and held high positions in Portugal. When he returned to Brazil in 1819, he played a crucial role in Brazilian independence.

The Case of New Spain

In the geography of Latin American science during the Enlightenment there were differences and similarities. In the case of New Spain, the primary viceroyalty from the economic point of view, the general aspects of the Hispanic American Enlightenment were certainly present. I mention New Spain separately because some of these characteristics were more structured and executed more widely as a consequence of the social dynamics there, as how science was organized, for example. In other places, scientific activities were organized in significant ways that suggest correspondence with the process in New Spain (I do not detail them here for lack of space).

Indeed, the incorporation of modern science into the northern part of Hispanic America resulted from the uninterrupted work of several generations of scientists in New Spain who, acting as a scientific community, created complex ways to organize their activity. In the first half of the eighteenth century, geography, astronomy, medicine, metallurgy, and botany, as well as industrial arts and technology, were still cultivated in

Mexico in individual modalities (as opposed to the institutionalized ones established later). It should be noted, however, that scientists no longer worked only in the capital of the viceroyalty; rather, their activities and influence extended to other parts of the territory, and different cities, like Mérida, Puebla, Valladolid, and Zacatecas, had groups of scientists, publications, and institutions that fostered their work. The scientific works of this period contain elements of both the modern and the traditional; that is, the gradual introduction of mechanist philosophy, the latest product of European science, can be observed in them. Some of the scientists who stood out (and who were, for all intents and purposes, self-taught) were José Sáenz de Escobar, Francisco Javier Alejo, José de Rivera Bernaldez, and Narciso Macop.

In the first half of the century, inventions and applied science marked progress in areas such as the crafting of musical instruments, bells, and machines for extinguishing fires; in agriculture (wheat and sugarcane milling, tobacco growing and processing, etc.); and in the mechanical arts (the construction of mills to process minerals, ovens and lathes for the minting of money, metals processing, mine drainage, textile and gunpowder production, and pottery making, etc.). This shows how developed practical knowledge was regarding improving procedures, instruments, apparatuses, and devices—even within the limitations imposed by the monopoly Spain maintained over most industrial products.

Institutions like the Mexico City mint (established in 1732) were very important in the development of the industrial arts. The great amounts of silver and gold minted there necessitated continual improvements in procedures and methods.[40]

An interesting aspect of this period is the polemic concerning the value and dimension of the Mexican intellectual tradition, which was a factor in reinforcing the nationalist feeling of the Creole elite of New Spain. The case of Juan José de Eguiara y Eguren constitutes a brilliant example. He reacted to the contempt and defamation the dean of the church of Alicante, Manuel Martí, heaped on Mexico by publishing in 1755 the learned work *Bibliotheca mexicana* (Mexican Bibliography). It includes a catalog of the numerous publications that had appeared in the country since the sixteenth century, when the printing press was introduced.

In the second half of the eighteenth century, scientific activity in New Spain grew considerably in terms of quantity and quality: an active scientific community was created, which had the resolute support of various sectors of society; its members displayed a vast, truly encyclopedic scientific culture and interest in areas that, at that time, were on the "frontiers" of science; the scientific community meshed its activities with technical, productive, governmental, cultural, ideological, and political action; science and technology in research and lay teaching institutions was institutionalized, supported completely or partly by the people of New Spain; interest grew in popularizing science, education, and the useful arts as elements of a social reform program that included the formation of scientific culture in the country; a strong nationalism sprang up and created an interest in knowing the country and the provinces, their resources, and their history; new professional relationships developed among scientists, the people, and institutions from other European and American countries. All these conditions led the Enlightenment science of New Spain to acquire its own characteristics, different from the European scientific mold, because New Spain was not seeing simply the spread or transference of science and institutions to Mexican ground but the transfusion or adoption of science. This was the moment when science first began to play a true role in New Spain.

The scientific community that formed in this period came from a very important sixteenth-century background. What is particular to the times was the breaking up of isolation and social marginality as this community established alliances with various sectors of New Spain's society—now interested in modern science and willing to support it. These associations led to the growth of the community and the expansion of its influence. Also distinctive was the development and consolidation of an ideology—scientific nationalism—that unified scientists and allowed them to achieve higher standing in society. In fact, both aspects are intermingled and mutually associated.

The central role of science in New Spain at the end of the eighteenth century, its attachment to the customs, values, and idiosyncrasies of society, was the result of adopting science (to "tame" it). We can classify the factors that determined this process into four groups: mining and other economic activities; public works; culture and education; and the exploration of the territory, its natural resources, and its inhabitants. As we know, the interest in improving techniques related to mining and the processing of metals was long-standing and vital for the economy, and there were many achievements in this area throughout the colonial period.[41] Although this concern was important and helped improve mining, knowledge was not systematized. Systematization could have enabled scientists and mine owners to deal more efficiently with complex problems.

Furthermore, mining faced numerous bureaucratic obstacles; financing problems; lack of supplies and the high cost of quicksilver, gunpowder, and other inputs; excessive taxes and tributes; and old and enormously complex legislation. All of these problems hindered mining. Mine owners lacked the means to overcome their numerous problems and an organization to represent them. The Crown was not interested in solving these problems, either, and used the Bourbon reforms to centralize and increase its control over Mexican silver. Thus, it was necessary to approach the problem as a complex whole, and only the interested parties (mine owners, merchants, scientists, and technicians) could devise a solution by acting as a group. When this happened, the decisive step toward solving their problems was taken.

In 1761, Judge Francisco Javier Gamboa drew up a political plan and an economic, scientific, and technical study for mining that flew in the face of the Bourbon reforms. Gamboa represented the interests of merchants and mine owners against the reform projects personified in José de Gálvez, visiting judge in New Spain and later minister of the Indies. Besides being a juridical and economic project for mining, Gamboa's *Comentarios a las ordenanzas de minas* (Comments on Mine Ordinances) includes a systematic and complete description of the patio process, revealing for the first time chemicometallurgical techniques that had never come to light and that only the practical processors had mastered. Gamboa demonstrated the superiority of this method and also—for the first time—expressed the intention to create a school for the teaching of science (physics, chemistry, mathematics, etc.) for miners' children.[42] Gamboa's influence spread widely and over time because *Comentarios* was a necessary reference work for mining in various countries until well into the nineteenth century.[43]

In 1774, the brilliant mathematician and engineer Joaquín Velázquez Cárdenas de León and the mine owner Juan Lucas Lassaga, using Gamboa's *Comentarios*, took the next step by sending the king a "Representation" in which they again described the lamentable state of mining and suggested a set of financial, juridical, organizational, scientific, and technological reforms. One of these was the creation of a school of specialized instruction for miners' children and a program of modern scientific studies to create university-trained experts and professional scientists in mining and metallurgy. In 1777, they were successful in creating the Tribunal de Minería and, in 1783, received the approval of mining ordinances that incorporated the suggestions of the 1774 Representation and Gamboa's *Comentarios*. Those documents regulated all aspects of mining and, most important for our study, authorized the creation of a seminar or school of mining, although the training of scientists was excluded.

The Real Seminario de Minería (Royal Mining College) began in 1792, under the direction of Spanish mineralogist Fausto de Elhúyar. His appointment (which violated regulations and displaced the Creole scientists who had conceived the institution and requested the position) and the appointment of European professors of mathematics, physics, chemistry, and mineralogy also created uneasiness and justified protests from the country's most eminent scientists, such as mathematician and astronomer Antonio de León y Gama and J. A. Alzate. Likewise, there were important polemics about the metallurgical methods Elhúyar sought to introduce (in 1791), most notably, the arguments against them formulated by Alzate and their defense by Francisco Javier de Sarria using Lavoisian nomenclature.

The regular teaching of several modern sciences with applications for mining began in this institution, and its existence represented a victory for the local scientific community. For this reason, local scientists worked for the Real Seminario de Minería, which was vital to its operation and the achievement of its aims. The college catalyzed the important movement in New Spain that comprised a scientific community made up of a considerable number of individuals, a tradition, publications, libraries, instruments, mineral collections, and awareness of the possibilities and importance of the movement. Its central role in New Spain became a social and economic reality, too. Graduates and those attending classes quickly became the palpable proof of the role science was to play in society when they went to work in the mines and in other activities.

Upheaval in the academic life of the college was caused by a series of events: the new king (Charles IV) was indifferent toward science and feared the potential effects of the Enlightenment and the example of the French Revolution on his domain; the European wars hindered communications with America, so the supply of books, instruments, mineral collections, and professors from Spain was delayed, too; the Real Seminario de Minería suffered financial problems as a consequence of inefficient management, debts, and loans to the king that reduced its funds. Surely, even one of these causes would have had been enough for the project to miscarry. Quite the opposite happened, however, because of the participation of the local scientific community and other social sectors. The college continued to operate and accomplished its proposed goals and others not initially contemplated, but important to New Spain. The support of mine owners rectified the institution's financial problems; the support of scientists teaching in the college and in other institutions made up for the initial lack of professors as graduates took charge of the teaching of several subjects. Locals began to translate books and, most important, began to prepare texts specially designed for teaching at the Seminar. Especially significant was the translation (the first into Spanish) of the first volume of *Traité élémentaire de chimie* (Elementary Treatise on Chemistry), by Lavoisier, and the writing of a course on modern physics as applied to mining problems, by Francisco Antonio Bataller. This text, *Principios de física y matemática experimental* (Principles of Physics and Experimental Mathematics, 1802), is remarkable because it was the first text on proper Newtonian physics written in the country and according to a practical concept: physics adapted for application to specific mining problems in New Spain (mechanics, hydraulics, aerodynamics, etc.). Earlier texts on modern physics written by authors in New Spain, such as those by the Jesuit Francisco Javier Clavijero (*Physica particularis* [Special Physics] 1765) and the Phillipian Juan Benito Díaz de Gamarra (*Elementa recentioris philosophiae* [Elements of Modern Philosophy] 1774), were explanatory, immersed in philosophical debates and with a rather eclectic focus.[44] Science, however, now had clearly practical functions to carry out, and it had fully entered into a phase of useful applications. It was also an instrument for the education of a new type of professional in the Hispanic American environment: the engineer.

The Real Seminario de Minería relied on eminent scientists such as the Spaniard (and, later, naturalized Mexican) Andrés Manuel del Río, who wrote *Tratado de orictognosia* (Treatise on Oryctognosy) to be used by students; he also discovered the chemical element called "erythronium" (vanadium).[45] German scientists also contributed: Ludwig Lindner, taught chemistry; Friedrich Sonneschmidt, wrote a chemical and mineralogical study in German translated as *Tratado de la amalgamación en México* (Treatise on Amalgamation in Mexico, 1805) in which he recognizes the superiority of the patio process and recommends its use in Europe. In 1802, José Garcés y Eguía had written *Nueva teórica y práctica del beneficio de los metales* (New Theory and Practice of Metals Processing), and an anonymous author had written *Tratado de docimacia* (Treatise on Docimasy); both works explain the traditional metallurgy method (the patio process) by means of modern chemistry.[46] In 1803, Alexander von Humboldt visited the college for several months, taught classes, and spoke well of the institution.

We can include the Academia de San Carlos in what I am calling centers of economic and public works. It was recognized by the king of Spain in 1785, but had, in fact, been operating for several years as a result of initiative taken by and the financial and scientific support of society.[47] This academy was created in response to a need for trained engravers for the mint, but very soon the usefulness of offering scientific and technical studies aimed at improving factories and public works became obvious. This institution taught artisans and architects in Mexico City (where many of them gathered) mathematics, drawing, engraving, painting, sculpture, architecture, and engineering. Eventually, more than three hundred students passed through the academy's doors. José Ignacio Bartolache was its secretary, and Diego de Guadalajara (a teacher and textbook author), both eminent mathematicians, stand out among its professors, as do advisers Joaquín Velázquez Cárdenas de León (mathematician and astronomer) and Fausto de Elhúyar (mineralogist), academy president Gerónimo Gil, and professors and architects Miguel Constanzó and Manuel Tolsá. Some of the academy's brilliant graduates included architects José Damián Ortiz de Castro, Ignacio Castera, and José Antonio González Velázquez. Professors and graduates of the academy (and the academy itself as responsible for authorizing construction work) designed and constructed fortifications, buildings, aqueducts, roads, pavement, drains, public parks, and other public works for sanitary engineering; on the whole, these contributed important solutions to the social and economic life of the country and to the comfort of its inhabitants. In the case of Mexico City, these public works gave it the reputation as the city of palaces.

In New Spain, which was part of the "education and culture" hub, scientists participated in journalistic endeavors uninterruptedly starting in the 1770s, as previously mentioned. The appearance of scientific journalism and its permanence, in spite of censorship and the viceregal government's prohibitions, reveals the widespread interest in "enlightenment," the emergence of a new mentality among the elite, and, more specifically, the existence of a scientific tribune in which to publish new theories (like those of Lavoisier, Linnaeus, Newton, etc.) and even to debate innovation (Linnaeus, Lavoisier, Bohr).

In the education sphere, the appearance in the 1780s and 1790s of new institutions with a modern bent, such as the Real Seminario de Minería, the Academia de San Carlos, and the Jardín Botánico and Cátedra de Botánica, brought into being the Enlightenment ideal of making science an instrument in the struggle against ignorance, fanaticism, and superstition in addition to a way to attain material progress. Prior to the creation of these institutions, some religious schools (run mainly by Jesuits) taught science in the brief moments when there was a certain amount of freedom (as in Morelia and Querétaro, for example), which meant that those who had traveled to Europe or were self-taught could teach modern philosophy and science from an eclectic perspective. Nevertheless, repression imposed by the religious orders themselves (as in the cases of José Pérez Calama and Benito Díaz de Gamarra),[48] the expulsion of the Jesuits from the viceroyalty in 1767, and, primarily the lack of social support for these initiatives, prevented these attempts from institutionalizing the teaching of science.

Once the teaching of science was institutionalized in lay institutions, its biggest impact was in the cultural and ideological areas. The useful arts, which guided the Creole reform program, saw only scant results as a consequence of inconsistencies that resulted from colonial ties and increased under the administrative centralization and authoritarianism reforms introduced by the Bourbons.[49] Regarding mining, for example, there were few cases in which there was true technical improvement. The nature of mining, and other administrative and political reasons, made proprietors consider technical matters their exclusive concern and reject the innovations Elhúyar arrogantly proposed in his role as director of the Real Seminario de Minería—an attitude for which he was reproached by mine owners.[50]

Meanwhile, these same institutions, taking advantage of conditions created by Creoles over several decades of intense scientific and popularizing activity, stimulated the appetite of New Spain's elite for modernity. In the space of a few years, a generation of scientific and learned men from different professions (e.g., medicine, pharmacy, law, architecture, the church) were trained and attended scientific lectures as "devotees," "boarding-school students," or practitioners. Some of them stood out in the cultural life of the colony, for example, José Mariano Mociño in natural science; José Luis Montaña (1755–1820) in medicine and chemistry; Manuel Cotero in chemistry. This cultural effect (even though not sought by officials) linked European science and Enlightenment thought and a great number of Mexicans. It was very important in New Spain at the end of the colonial period and an important factor in consolidating an ideology of independence. I am not overstating when I say that a good many scientists who espoused Enlightenment ideals participated in New Spain's War of Independence (1810–1821) and died alongside the rebels.

Finally, "knowledge of the territory and its natural and human wealth" marked Hispanic American Enlightenment nationalism. This telluric feeling that ties men born in a certain territory to their environment, or even those lately arrived, as was the case of many Europeans who became naturalized Americans (e.g., the Bohemian Leopold Hancke in Charcas; the Spaniard Vicente Cervantes in Mexico; José Celestino Mutis in New Granada; the Portuguese Antonio Parra in Cuba), is at the heart of the important scientific work carried out in the Americas in botany, zoology, paleontology, mineralogy, geology, in what made man and society American, in anthropology, archaeology, linguistics, and history.

This interest in the territory and its inhabitants was twofold. On the one hand, it was simply knowledge, an unavoidable wish to know before the immediate and common reality of Americans. But it was not usually a part of established (i.e., European) science, which either ignored it or scorned it to a degree that in effect established the inferiority of Hispanic American nature, man, and society (see, e.g., De Pauw, Robertson, Buffon, and Manuel Martí).

On the other hand, there was a pragmatic purpose in studying the territory and its inhabitants: to exploit resources and use them to help the "patria del criollo" (commonwealth of the Creole),[51] since the socioeconomic and cultural interests of other sectors of society such as Indians, mestizos, and mulattoes were not being taken into consideration by the ruling classes.[52]

For both purposes, it was very important that cartography, observations of position astronomy and astronomical phenomena, travels and exploratory expeditions, descriptions of fauna and flora, study of plants and their classification, mineral collections and prospecting for energy resources, and study of disease, among other activities, be carried out by scientists in New Spain; all of these activities led to detailed knowledge of the land and its resources. In the same way, in the humanities, studies and collection of archaeological objects were important, as were dictionaries of indigenous languages. Descriptions of the customs, religious beliefs, and native ways of life; indigenous chronology; and the history of institutions and culture in New Spain were all equally vital. These types of studies were carried out by individuals steeped in Enlightenment ideals and nationalist ideology; they were frequently the same people interested in the study of nature.

The creation of the Jardín Botánico (in 1788) as part of the Expedición Botánica to New Spain (proposed from Mexico by Spanish physician Martín Sessé) was a very important initiative for investigating the flora of the country and was coupled with the teaching for the first time of Lavoisian chemistry in the Jardín Botánico.[53] Moreover, the Jardín Botánico contributed to reforms in teaching medicine and pharmacy in New Spain.[54]

At the beginning of all these initiatives, they suffered from the characteristic authoritarianism of the Crown and its search for exclusive benefits—which embroiled the Jardín Botánico, the Expedición Botánica, and the Cátedra de Botánica in serious confrontations with the university, the Protomedicato (Board of Medical Examiners), and with individuals like Alzate regarding juridical, organizational, financial, and theoretical issues.[55] The expedition, the garden, and the department, however, eventually were fully incorporated into the scientific life of New Spain.[56] The department gave modern instruction to physicians, pharmacists, and many "inquisitives." The expedition traveled throughout New Spain and Guatemala, carrying out significant work in description and taxonomy as well as collecting specimens of Mexican flora.

Distinguished members of the scientific community collaborated in and contributed to this work, among them, physician José Mariano Mociño of New Spain. Besides participating in the expedition's main outings, this scientist visited and described the botanical, anthropological, and linguistic aspects of Nutka, on the northwestern coast of the continent (present-day Vancouver). Later, in Madrid, he was responsible for the organization for publication of the materials collected by the expedition, and between 1808 and 1812, he directed the Gabinete de Historia Natural (Office of Natural History). He also presided over the Academia de Medicina and researched yellow fever after which he wrote a report.[57] Mociño thus became the third American to hold a highly responsible position in Spain's scientific organizations; he was preceded by Peruvian Francisco Dávila, the first director of the Gabinete de Historia Natural, and by Francisco Antonio Zea, from Medellín, a pupil of Mutis, who was appointed director of the Jardín Botánico of Madrid in 1804.

Conclusions

In a little more than one hundred years, regional historical dynamics led to a fundamental change: the transformation of Spanish America into Latin America. Indeed, as we have seen, the constant growth of the mining, agricultural, and handicrafts economy beginning in approximately 1700, as well as the presence of a social and racial structure in which Creoles (being the only ones able to create wealth, an endogenous cultural movement, and a nationalist ideology that agglutinated the different social groups) were the leaders, modified the bases of the Iberian-American colonial regime and led to a decisive crisis. In its place, a group of societies began to appear that acquired an increasingly higher degree of autonomy in practically all areas as well as an awareness of themselves.

The Latin American nations, all of them possessing similar material, racial, and cultural conditions and a similar historical process, acquired characteristics that, though they were identified, remained hidden from protagonists. To reveal these characteristics, the Latin Americans incorporated and adapted science and Enlightenment ideals as cognitive practices and ideology, respectively, thereby enabling themselves to supply the cultural answer to an unheard-of situation: the emergence of the geocultural and historical entity that is today Latin America.

It was up to the Latin Americans themselves to invent appropriate solutions to their problems by using their own resources; no behavioral formulas existed that could be applied to their reality.[58] Science was validated socially through difficult negotiations among various sectors and via initiatives undertaken by the intellectual elite. When scientists found allies interested in cultural, economic, and political modernization, they incorporated the Enlightenment ideal of social reform into their practices; this made it necessary to adapt European science. This was the only way to get beyond the level of learned scientific culture, individually or in small groups, and to get science institutionalized and socially recognized.

As a consequence, the incorporation of modern science into American society took place in a local context, which led scientific practice to acquire its own style. Additionally, at the end of the eighteenth century and the beginning of the nineteenth, science took on a leading role in social transformation of the region, and it became one of the cultural and material agents of change.

Finally, the incorporation of modern science into Latin American society took place when a new social framework was being created (in which science was fully incorporated), a framework that did not have anything to do with the authoritarian and colonial political regime that had governed until then. In their way, Latin American scientists also struggled for freedom and independence, the only framework in which science could develop and fulfill a social function. Latin American scientific Enlightenment was, therefore, the achievement of societies in the process of transformation and in search of their identity.

Notes

1. Brading, *Mineros*.
2. Florescano and Gil, "La época." [In Mexico, the word *Bajío* ("low-lying ground") means a very specific zone: the State of Guanajuato and some areas along the border states—Trans.]
3. Brading, *Mineros*; Condarco, *Historia*; Garavaglia, *Mercado interno*; Molina, *El Real Tribunal*.

4. Sánchez, *Historia*.
5. Howe, *The Mining Guild*; Molina, *El Real Tribunal*.
6. Walker, *Política española*.
7. Luque, *La sociedad económica*.
8. Hankins, *Ciencia e Ilustración*.
9. Saldaña, "The Failed Search."
10. Saldaña, *Las fases principales*.
11. Rodríguez, *Historia*.
12. Vilchis, "Medicina novohispana."
13. Osorio, *Historia*.
14. Barreda, *Vida intelectual*.
15. Osorio, *Historia*.
16. López, "Estrategias comerciales."
17. Romero, "El Mercurio Peruano"; Clement, "El surgimiento."
18. Osorio, *Historia*.
19. Bateman, *Francisco José de Caldas*.
20. Osorio, *Historia*.
21. Águila, "El periodismo científico."
22. Albornoz, "Eugenio Espejo"; Estrella, *José Mejía*; Saladino, "La ciencia."
23. Molina, *El Real Tribunal*; Saldaña, "The Failed Search."
24. Arboleda, "Science and Nationalism"; Saldaña, "Nacionalismo."
25. Arboleda, "Acerca del problema."
26. Barreda, *Vida intelectual*.
27. Trabulse, "Tres momentos."
28. Trabulse, "La ciencia perdida."
29. Trabulse, "La obra científica."
30. Vilaseca, "Matemáticas y astronomía."
31. Trabulse, "La obra científica."
32. Saldaña, "Nacionalismo."
33. Hernández de Alba, *Pensamiento científico*.
34. Arboleda, "Sobre una traducción inédita."
35. Arboleda, "La ciencia."
36. Lastres, "El pensamiento científico-natural."
37. Barreda, *Vida intelectual*.
38. Lastres, *Hipólito Unánue*.
39. Díaz, "La ciencia moderna."
40. Sánchez, *Historia*.
41. Bargalló, *La minería*.
42. Saldaña, "The Failed Search."
43. Trabulse, *Francisco Xavier Gamboa*.
44. Izquierdo, *La primera casa*; Navarro, *Cultura mexicana moderna*; Ramos, "La difusión."
45. Rubinovich, "Andrés Manuel del Río."
46. Bargalló, *La minería*.
47. Brown, *La Academia de San Carlos*.
48. Cardozo, *Michoacán*; Jaramillo, *José Pérez Calama*.
49. Florescano and Gil, "La época."

50. Saldaña, "The Failed Search."
51. Martínez, *La patria del criollo*.
52. Martínez, 1982.
53. Trabulse, "Aspectos."
54. Islas and Sánchez, *Historia*.
55. Tanck, "Justas florales."
56. Zamudio, "Institucionalización."
57. Dívito, "Mociño."
58. Chenu, "Desde la tierra."

Bibliography

Águila, Y. "El periodismo científico en Nueva España: Alzate y Bartolache (1768–1773)." In *La América española en la época de las luces*. Madrid: Ediciones de Cultura Hispánica, 1988.

Albornoz, M. "Eugenio Espejo, médico de Quito del siglo XVIII y hombre de ciencia." In *Memorias del Primer Coloquio Mexicano de Historia de la Ciencia*, vol. 2. Mexico City, 1964.

Arboleda, L. C. "Acerca del problema de la difusión científica en la periferia: El caso de la física newtoniana en la Nueva Granada." *Quipu, Revista Latinoamericana de Historia de las Ciencias y la Tecnología* 4, no. 1 (1987): 7–30.

———. "La ciencia y el ideal de ascenso social de los criollos en el Virreinato de Nueva Granada." In *Ciencia, técnica y estado en la España ilustrada*, J. Fernández and I. González (eds.). Madrid: Ministerio de Educación y Ciencia, 1990.

———. "Science and Nationalism in New Granada on the Eve of the Revolution of Independence." In *Science and Empires: Historical Studies about Scientific Development and European Expansion*, P. Petitjean and C. Jami (eds.). Dordrecht: Kluwer Academic Publishers, 1991 and 1992.

———. "Sobre una traducción inédita de los *Principia* al castellano hecha por Mutis en la Nueva Granada *circa* 1770." *Quipu, Revista Latinoamericana de Historia de las Ciencias y la Tecnología* 4, no. 2 (1987): 291–314.

Babini, J. *La evolución del pensamiento científico en la Argentina*. Buenos Aires: La Fragua, 1954.

Bargalló, M. *La amalgamación de los minerales de plata en Hispanoamérica colonial*. Mexico City: Compañía Fundidora de Fierro y Acero de Monterrey, 1969.

———. *La minería y la metalurgia en la América española durante la época colonial*. Mexico City: Fondo de Cultura Económica, 1955.

Barreda, L. F. *Vida intelectual del Virreinato del Perú*. Lima: Universidad Nacional Mayor de San Marcos, 1964.

Bateman, A. D. *Francisco José de Caldas: El hombre y el sabio*. Cali: Biblioteca Banco Popular, 1978.

Bosch, C. *La polarización regalista de la Nueva España*. Mexico City: Universidad Nacional Autónoma de México, 1990.

Brading, D. A. *Mineros y comerciantes en el México borbónico (1763–1810)*. Mexico City: FCE, 1975.

Brown, T. *La Academia de San Carlos de la Nueva España*. Mexico City: Sepsetentas, 1976.

Cardozo, G. *Michoacán en el siglo de las luces*. Mexico City: El Colegio de México, 1973.

Chenu, J. "Desde la tierra hacia las estrellas: Búsqueda científica e identidad cultural en Nueva Granada." In *La América española en la época de las luces*. Madrid: Ediciones de Cultura Hispánica, 1988.

Clement, J. P. "El surgimiento de la prensa periódica en la América española: El caso del 'Mercurio Peruano.'" In *La América española en la época de las luces*. Madrid: Ediciones de Cultura Hispánica, 1988.

Condarco, R. *Historia de la ciencia en Bolivia*. La Paz: Academia Nacional de Ciencia, 1978.

Díaz, L. "La ciencia moderna en Cuba a principios del siglo XIX: Las fuentes de la física de Félix Varela." *Asclepio* 42 (1990): 393–402.

Dívito, J. C. "Mociño y la fiebre amarilla." *Historia Mexicana*, 15, no. 1 (1964): 1–27.

Duque, L. M. "Presencia de las ideas ilustradas acerca de las ciencias en el pensamiento neogranadino de fines del siglo XVIII: El caso de Francisco José de Caldas." MA thesis, Universidad Nacional Autónoma de México, 1988.

Estrella, E. *José Mejía, primer botánico ecuatoriano*. Quito: Ediciones ABYA-YALA, 1988.

Ferry, M., and S. Motoyama (eds.). *História das ciências no Brasil*. 3 vols. São Paulo: EDUSP, 1979.

Florescano, E., and I. Gil S. "La época de las reformas borbónicas y el crecimiento económico." In *Historia general de México*, vol. 2, D. Cosío Villegas (ed.). Mexico City: El Colegio de México, 1976.

Garavaglia, J. C. *Mercado interno y economía colonial*. Mexico City: Grijalbo, 1983.

Hankins, T. *Ciencia e Ilustración*. Madrid: Siglo XXI, 1988.

Hernández de Alba, G. (ed.). *Pensamiento científico y filosófico de José Celestino Mutis*. Bogotá: Ediciones Fondo Cultural Cafetero, 1982.

Howe, W. *The Mining Guild of New Spain and Its Tribunal General, 1770–1821*. New York: Greenwood Press, 1958.

Islas, V., and J. F. Sánchez. *Historia de la farmacia en México y en el mundo*. Mexico City: Asociación Farmacéutica Mexicana, 1992.

Izquierdo, J. J. *La primera casa de las ciencias en México*. Mexico City: Ciencia, 1955.

Jaramillo, J. *José Pérez Calama, un clérigo ilustrado del siglo XVIII en la antigua Valladolid de Michoacán*. Morelia: Universidad Michoacana de San Nicolás de Hidalgo, 1990.

Lastres, J. B. *Hipólito Unánue*. Lima: PGACE, 1955.

———. "El pensamiento científico-natural en el Perú a fines del siglo XVIII." *Revista Universitaria* (Cuzco) (1953): 105.

López, F. "Estrategias comerciales y difusión de las ideas: Las obras francesas en el mundo hispánico e hispanoamericano en la época de las luces." In *La América española en la época de las luces*. Madrid: Ediciones de Cultura Hispánica, 1988.

Luque, E. *La educación en Nueva España*. Seville: Escuela de Estudios Hispanoamericanos de Sevilla, 1970.

———. *La Sociedad Económica de Amigos del País de Guatemala*. Seville: Escuela de Estudios Hispanoamericanos de Sevilla, 1962.
Marco, D. E. *Materiales para la historia de la cultura en Venezuela, 1523–1828*. Caracas: Fundación John Boulton, 1967.
Martínez, S. *La patria del criollo: Ensayo de interpretación de la realidad colonial guatemalteca*. Puebla: Universidad Autónoma de Puebla, 1982.
Molina, M. *El Real Tribunal de Minería de Lima (1785–1821)*. Seville: Diputación Provincial de Sevilla, 1986.
Motten, C. G. *Mexican Silver and the Enlightenment*. New York: Octagon Books, 1972.
Navarro, B. *Cultura mexicana moderna*. Mexico City: Universidad Nacional Autónoma de México, 1982.
Osorio, I. *Historia de las bibliotecas novohispanas*. Mexico City: Secretaría de Educación Pública, 1986.
Paz, C. E. *Hipólito Unánue: El padre de la medicina americana*. Lima: Talleres Gráficos del Asilo Víctor Larco Herrera, 1925.
Pérez, J. "Tradición e innovación en la América del siglo XVIII." In *La América española en la época de las luces*. Madrid: Ediciones de Cultura Hispánica, 1988.
Peset, J. L., and A. Lafuente (eds.). *Carlos III y la ciencia de la Ilustración*. Madrid: Alianza Universidad, 1988.
Ramos Lara, M. de la Paz. "La difusión de la mecánica newtoniana en la Nueva España." MA thesis, Universidad Nacional Autónoma de México, 1991.
Rodríguez, A. M. *Historia de las universidades hispanoamericanas*. 2 vols. Bogotá: Instituto Caro y Cuervo, 1973.
Romero, E. "*El Mercurio Peruano* y los ilustrados limeños." In *Memorias del Primer Coloquio Mexicano de Historia de la Ciencia*, E. Beltrán (ed.). Vol. 2. Mexico City, 1964.
Rubinovich, R. "Andrés Manuel del Río y sus elementos de orictognosia de 1795–1805." In *Elementos de orictognosia, 1795–1805*, by Andrés Manuel del Río, R. Rubinovich (ed. and intro.). Mexico City: Universidad Nacional Autónoma de México, 1992.
Saladino, A. "La ciencia entre los ilustrados del Nuevo Mundo." PhD dissertation, Universidad Nacional Autónoma de México, 1988.
Saldaña, J. J. "The Failed Search for 'Useful Knowledge': Enlightened Scientific and Technological Policies in New Spain." In *Cross Cultural Diffusion of Science: Latin America*, J. J. Saldaña (ed.). Cuadernos de Quipu 2. Mexico City: Sociedad Latinoamericana de Historia de las Ciencias y la Tecnología, 1987.
———. "Las fases principales de la evolución de la historia de la ciencia." In *Introducción a la teoría de la historia de las ciencias*, 2nd. ed., J. J. Saldaña (comp.). Mexico City: Universidad Nacional Autónoma de México, 1989.
———. "Nacionalismo y ciencia ilustrada en América." In *Ciencia, técnica y estado en la España ilustrada*. Madrid: Ministerio de Educación y Ciencia, 1990.
Sánchez, R. *Historia de la tecnología y la invención en México*. Mexico City: Fomento Cultural BANAMEX, 1980.
Soto, A. D. *Las polémicas universitarias en Santa Fe de Bogotá: Siglo XVIII*. Bogotá: Colciencias-UPN, 1993.

Tanck de Estrada, D. "Justas florales de los botánicos ilustrados." *Diálogos* 106 (Mexico) (1982).

———. "Tensión en la torre de marfil: La educación en la segunda mitad del siglo XVIII mexicano." In *Ensayos sobre historia de la educación en México*. Mexico City: El Colegio de México, 1985.

Trabulse, E. "Aspectos de la difusión del materialismo científico de la Ilustración francesa en México a principios del siglo XIX." *Quipu, Revista Latinoamericana de Historia de las Ciencias y la Tecnología* 6, no. 3 (1989): 371–386.

———."Aspectos de la tecnología minera en Nueva España a finales del siglo XVIII." *Historia Mexicana* 30, no. 3 (1981): 311–357.

———. *La ciencia perdida: Fray Diego Rodríguez, un sabio del siglo XVII*. Mexico City: FCE, 1985.

———. *Ciencia y religión en el siglo XVII*. Mexico City: El Colegio de México, 1974.

———. *Francisco Xavier Gamboa: Un político criollo en la Ilustración mexicana*. Mexico City: El Colegio de México, 1985.

———. "Introduction." In *Historia de la ciencia en México*. vol. 3. Mexico City: FCE, 1983.

———."La obra científica de don Carlos de Sigüenza y Góngora (1645–1700.)" In *Actas de la Sociedad Mexicana de Historia de la Ciencia y de la Tecnología*, V. González (ed.). Vol. 1. Mexico City, 1989.

———. "Tres momentos de la heterodoxia científica en el México colonial." *Quipu, Revista Latinoamericana de Historia de las Ciencias y la Tecnología* 5, no. 1 (1988): 7–18.

Vilaseca, S. "Matemáticas y astronomía en la historia de Cuba." *Quipu, Revista Latinoamericana de Historia de las Ciencias y la Tecnología* 2, no. 2 (1985): 185–212.

Vilchis, J. "Medicina novohispana del siglo XVI y la materia médica indígena: Hacia una caracterización de su ideología." *Quipu, Revista Latinoamericana de Historia de las Ciencias y la Tecnología* 5, no. 1 (1988): 19–48.

Walker, G. *Política española y comercio colonial, 1700–1789*. Barcelona, 1979.

Worcester, D., and W. Schaeffer. *The Growth and Culture of Latin America: From Conquest to Independence*. Vol. 1. Oxford: Oxford University Press, 1970.

Yepes, E. (ed.). *Estudios de historia de la ciencia en el Perú*. Vol. 1. Lima: CONCYTEC, 1986.

Zamudio Varela, G. "Institucionalización de la enseñanza de la botánica en México (1787–1821)." MA thesis, Universidad Nacional Autónoma de México, 1991.

CHAPTER 3

Modern Scientific Thought in Santa Fe, Quito, and Caracas, 1736–1803

LUIS CARLOS ARBOLEDA AND DIANA SOTO ARANGO

Introduction

The Enlightenment current that permeated the Viceroyalty of New Granada in the late eighteenth century—simultaneous with the reign of Carlos III—encouraged a new, "useful," philosophy, which involved the teaching of Newtonian natural philosophy. The theories and concepts of this new philosophy were aimed at opposing Scholastic thought by explaining reality through observation and experience. They were also an attempt to promote a different vision of the world in new generations by teaching modern physics in nontraditional ways.

Before they were incorporated into programs of study at the universities of New Granada, Copernican theses could be found, for example, in the activities of the scientific missions, which carried out geodetic measurements, settled territorial limits and borders, or built fortifications. Among the most famous expeditions financed by the Bourbons was Charles de la Condamine's geodetic mission to the Viceroyalty of Peru, which included Spanish scientists such as Antonio Ulloa and Jorge Juan. The expeditionaries arrived in the city of Quito around 1736 and fostered a favorable climate for discussing modern science in that city. The Universidad Gregoriana became a remarkable place for the development of the scientific spirit in the 1740s.

Santa Fe underwent similar changes, as the Universidad Javeriana was the first educational institution that permitted the teaching of Enlightenment philosophy. The Universidad de Caracas, however, did not allow the introduction of Copernicus's and Newton's theories until later, by Prof. Baltasar de los Reyes Marrero, around 1788.

The nearly half century–long polemic on Copernican theories, which started in 1773 between José Celestino Mutis and the Dominican Congregation of Santa Fe de Bogotá, was very important. This controversy involved not only rival philosophical concepts but also different social and political interests and drew in two organized groups aiming to dominate education. The attempts to introduce and develop Enlightenment philosophy in the universities of Santa Fe, Quito, and Caracas came from within the academic environment. Professors like Father Juan de Hospital, Miguel Antonio Rodríguez, José Mejía Lequerica, and Pedro de Quiñones in Quito, Marrero from the Universidad de Caracas, and, among others, Mutis, Valenzuela, Vallecilla, Vásquez, and Padilla, in Santa Fe's universities taught Newton's and Copernicus's theories in their philosophy or mathematics lessons. These professors attempted quite successfully to awaken scientific interest in the rigid university life of Santa Fe, Caracas, and Quito. Their efforts to institutionalize Enlightenment thought and scientific knowledge in university classrooms were remarkable, although, unfortunately, they were never fully successful during the colonial period.

In this chapter, we shall analyze facts relevant to the effort to introduce and defend modern scientific thought in Santa Fe, Quito, and Caracas. We shall keep in mind the different contextual circumstances of these cities regarding when Enlightenment philosophy was introduced, developed, and crystallized (with its Encyclopedist component).

The 1736–1803 period was characterized by the following:

1. 1736–1767. In 1736, La Condamine's expedition arrived in Quito; in 1767, the Jesuits were thrown out of Spanish territory. This stage is characterized by eclectic teaching by university staff and by the cultural effects in the viceroyalty of the geodetic expedition, especially in the City of Quito.
2. 1767–1783. In 1783, the Expedición Botánica of Santa Fe began work and was institutionalized. This period may be identified by, first, the debates about the inclusion of the new philosophy in college programs and, second, attempts to reform and secularize the universities.
3. 1783–1803. This was a period of state repression of Creole professors and students. Dominicans recovered their power and privileges in these institutions, and debates and gatherings were reported in the bulletins and newspapers of the time. Scientific activity and Newtonian theories were encouraged in the royal botanical gardens and the schools of mining and anatomy. The Sociedad Económica

(Economic Society) tried to inaugurate new ways of seeing and developing the economy of the colonies. In 1803, in the City of Quito, one of the last university debates to encourage Enlightenment thought in this institution [Quito's public university] was held.

The Bourbon Reforms

During the Bourbon administration, New Granada saw a series of reforms within Enlightenment despotism. The viceroyalty finally organized into a political unit in 1739. After several changes of name and administrative methods, the Real Audiencia de Caracas was founded in 1786. It should be kept in mind that the captain general of Venezuela exercised the same authority in his territory as did the viceroy of Santa Fe; therefore, they were absolutely independent of each other, and they dealt directly with the respective Spanish ministers.

The colony's basic economic sectors were mining, agriculture, and trade. The Viceroyalty of New Granada's income was only enough for its own subsistence. The survival of the viceroyalty depended on income from mining, especially of gold, but, although smuggling was prevalent in the gold trade, the greatest income came from taxes on rum. In Venezuela, "out of the 2,281,793 pesos' income collected," there was some left for the home country.[1]

Although the consolidation of the Creole and mestizo groups took place at the end of the century, political power was the aristocracy's monopoly. Creoles, in particular, "began to be aware of their historical initiative, influenced culturally by the Enlightenment movement and scientific awareness awakened by the Expedición Botánica."[2]

By the middle of the eighteenth century, university education in the Viceroyalty of New Granada was monopolized by the clergy, which can be easily explained by the weakness of both the viceregal state and the local civil elite. The latter could not establish secular institutions in which the new philosophy could be promoted. Besides, there was still a trend supporting traditional medieval education, which was used by the state-church coalition to subjugate the colonies.

After the Jesuits were banished, members of the Creole elite believed that the moment had arrived for them to direct and control education. They were vanquished by the power of the religious orders, however, which, paradoxically, were supported by the Crown. Quito's public university opened in 1788 and named a lay rector, Dr. Nicolás Carrión

Vacates, and three secular Creoles to give free classes at the university. In Madrid, however, it was decided to remove the rector and the three professors and to allow the bishop to name new professors.[3]

The case of Santa Fe's prosecutor (*fiscal*), Francisco Antonio Moreno y Escandón, is even more revealing about the power of religious communities. Moreno, in spite of his efforts to institutionalize new programs of study and a public university in Santa Fe, could not fully bring his idea to fruition. In New Granada, the reforming winds of colonial political regimes supported by the Creole elite blew in different directions. The state-church and state-military coalitions were not evident here until the early nineteenth century,[4] while in New Spain, for example, viceregal power had already agreed with local economic groups on several sociocultural projects, thereby creating institutions parallel to the university.

In the Viceroyalty of New Granada, the state (i.e., the non-Creole) academic elite coalition is seen only in the exceptional case of José Celestino Mutis, whose Expedición Botánica received extraordinary support starting in 1783. In contrast, we can see that an alliance was achieved between the Crown and the Creole elite in the Viceroyalties of Peru and New Spain outside the universities, for example, in the founding of the Escuela de Minas (School of Mining), the Jardín Botánico (Botanical Garden), and the Colegio de Cirugía (College of Surgery) in New Spain. Inside the universities, however, there continued a constant, eager debate caused by the incorporation of Enlightenment philosophy, as exemplified by the case of Prof. Ignacio Bartoloche.[5]

Thus, we may state an often forgotten fact: the cultural dynamics of the different viceroyalties cannot be studied without stressing the importance and economic priority of each. Furthermore, the wish of every religious order to monopolize education led them to fight each other, especially for the privilege of granting academic degrees. The Royal Edict of 1696, which established that both Jesuit and Dominican schools and universities could grant academic degrees, solved the problem in Quito; in Santa Fe, Philip V's Royal Decree of November 25, 1704, handled the situation in a similar manner.

In the Viceroyalty of New Granada, before the Jesuits were banished, debate concentrated in the communities of Santo Tomás and San Ignacio de Loyola. After 1767, the Dominicans were given the right to monopolize education. They requested donations, buildings, and prerogatives normally reserved for expatriates for the schools and universities of their community. In Santa Fe, we find the case of Father Jacinto Antonio Buenaventura, who, in 1777, confiscated all Jesuit property because of "how small their school [the Dominican] is" and "how little money they [the

Dominicans] have for acquiring science books." He also vowed to "keep the professorships and the [formerly Jesuit] Colegio Máximo with all its property, and the rights and privileges enjoyed by the old Universidad de los Jesuitas, along with all the prerogatives that Fiscal Moreno requested for the public university."[6]

By this time, however, the Crown had enacted its right of "royal patronage" and fully reestablished the Regium Execuator, which bequeathed to the state—under the king's dominion, not that of the pope—the control of public education and of university studies. The issue was never about free and obligatory education for the masses, or even directed at the working class, as we understand the term. Fiscal Moreno—a Creole reformer—said that, "if many, because of their poverty, do not have the necessary money to obtain a college degree and finance the customary pomp and gratuities, they will have to be content with a high school degree, which gives the necessary skills to employees, and the degree will be more appreciated and less common than at present."[7] On the contrary, during the years we are dealing with, university education was distinguished by its elite orientation to the Creole and Spanish elite destined to lead the local administration. Indeed, rigorous "information" was always an admission requirement for the university, and the cost of degrees rose.

Therefore, the Dominicans' attempt to keep the property of the banished Jesuits was thwarted by the decision of the Crown to control such property—and to keep it out of anyone's hands except the state's. For this purpose, ten Juntas de Temporalidades (Temporalities Boards) were formed throughout the Spanish kingdom, including the colonies. One of these, with the "charge of the viceroy of Santa Fe himself," belonged to the New Kingdom of Granada, and another was established in Caracas.

From the religious communities' standpoint, the importance of granting academic degrees through a specific institution guaranteed not only educational but, most important, political control. A review of relevant cases indicates that the Creoles who graduated from such institutions, because of the degree they received, occupied the few local administrative positions.

Although it is true that the Crown tried to reform and control education, the civil sector's attempts to direct university study in Santa Fe and Quito were actually subdued by the political-economic power of the clergy, specifically, by the Dominican order. After long debates in Santa Fe (and less-prolonged ones such as that held between the bishop of Quito and the rector and council of the Universidad de Quito in 1803), the Dominicans got the support of the Crown. They thus largely controlled university education and had the prerogatives and endowments of their respective schools: San Fernando in Quito and the Universidad Santo Tomás in Santa Fe. At the Universidad de Caracas, with its more secular tradition but under the wing of the Universidad Santo Tomás, Dominicans held two professorships starting in 1742.

Therefore, the hopes once held by the Creole elite, especially in Quito and Santa Fe, of controlling the universities were thwarted by traditionalist groups and by the Crown. The Crown had probably already detected in its American colonies hints of a nationalist project that relied on the new philosophy and therefore did not find any reason to institutionalize Enlightenment thought in universities.

Copernican and Newtonian Teaching in the Viceroyalty of New Granada

Before Copernican and Newtonian theories were instituted in universities in the Viceroyalty of New Granada, these theses had already circulated in a practical form in scientific missions to survey borders, to make astronomical observations, and to build fortifications. They might also have been disseminated institutionally at the Academia Militar de Matemáticas (Military Academy of Mathematics), established in Cartagena in 1731, at the Academia de Geometría (Academy of Geometry), and in the quarters created exclusively for officers in Caracas in 1760. These academies were created, among other reasons, to gather and compile information from travelers in order to stimulate interest in the natural sciences. One of the first visitors under the Bourbon regime was Amédie de Frezier, to the Viceroyalty of Peru, from 1712 to 1714. In 1735, sponsored by the French Académie Royale des Sciences and the Spanish Crown, the geodetic expedition led by Charles de la Condamine departed, taking along scientists like Pedro Bouguer, Luis Godin, Jean Seniergues, Antoine de Jussieu, and Spaniards Antonio Ulloa and Jorge Juan. Later, Hipólito Ruiz and José Pabón led an expedition to the Viceroyalty of Peru (1777–1788), and Sessé and Mociño led another one to New Spain (1788–1803). Between 1799 and 1804, Alexander von Humboldt and Aimé Bonpland moved around Spanish America under easier circumstances than had their predecessors. Certainly, La Condamine's expedition is considered "unfortunate" when compared to Maupertuis' expedition to Lapland at the same time to confirm Newton's assertion that the Earth was a sphere flattened on both poles; the latter took only eighteen months. Yet cultural outcomes—as Humboldt would point out years later—show that the interest in scientific investigation in these territories should be attributed, among other factors, to the influence exerted by Bourguer and La Condamine's expedition.[8]

The Jesuits, as primary sources show, were the first to systematically teach the theories of Descartes, Copernicus, and Newton at the universities of the Viceroyalty of New Granada.[9] It should be pointed out as well that by the middle of the eighteenth century, this community constituted an undeniable power, with the most significant influence on educational institutions in the entire world. This gave them, perhaps more than other orders, the possibility of internationalizing their educational strategies.

Being a true international force in culture and education, the order had to adapt its strategies to the new alliances in both fields as a way of maintaining its power. In the age of international capitalism, the new science had to be more pragmatic than speculative, and physics had to predominate over metaphysics. But adapting to these innovations was not an easy task for those who had legitimated the paradigm of systematic and Peripatetic science.

Despite the need to direct teaching toward change and modernity, tradition and old ways of thinking imposed an inertia, begetting a scaled-back modernity. Despite this, the Jesuits' power in the areas of culture and education and their internationalism and ability to mix with the elite and the court allowed them to increase their ability to innovate to the utmost and according to the scientific standards of the Enlightenment. This capacity radiated to the nodes of the international network through the different types of missions in which they participated and, in particular, through the order's official works.

Several Jesuit scholars moved throughout the colonies, carrying out scientific studies on American flora. Among others, we should mention José Cassani, who published his work on the flora of the Orinoco in Madrid in 1741; José Gumilla, who in the same year published "Orinoco ilustrado" (Orinoco Illustrated); and the Italian Jesuit Filipo Salvatore Gilij, who arrived in Santa Fe de Bogotá in 1743, where he lived for six years, to later travel the Meta and Orinoco rivers. He wrote his work on American vegetation using data collected on this trip.

Around 1730–1740, in France and Europe, the Jesuits switched from teaching according to the educational and cultural strategies of Descartes' mechanist and systematic philosophy (Descartes had earlier zealously refuted "Newtonian atheistic materialism") to teaching governed by strategies that sought to attune Cartesian philosophy to the principles of experimental physics. This shows the cultural adaptability of the order, similar to its most mundane political and social adaptation to the ruling powers in order to maintain its power behind the throne.

Leaving behind a defense of the Cartesian pole and coming to a Jesuit version of the Newtonian pole was not an automatic shift. It took several decades—catalyzed especially by the pressures of sociopolitical changes in the French and European context. The new theories, taught at the same time at the Universidad Javeriana of Santa Fe and the Universidad Gregoriana of Quito, were upheld by missionaries educated abroad or, in any event, aware of the transformations that were occurring in European institutions. The Universidad Gregoriana in Quito alone had "seventy-one foreign professors teaching at the university, and they recorded the subject in a manuscript. Native professors were twenty-one, of whom five were from Loja, four from Quito, three from Guayas, three from Cuenca, three from Riobamba, two from Ibarra, and one from Ambato."[10] This undoubtedly facilitated the exchange and dissemination of new ideas. These priests were familiar with and had experienced European intellectual reality. As a consequence, it is not strange that in a center of cultural ferment such as Quito, intellectual Jesuits were most closely linked to the Franco-Spanish geodetic mission directed by La Condamine and Jorge Juan.

A local Jesuit, Father Juan Magnin, established a close friendship with La Condamine. La Condamine had an educational and personal relationship with the Jesuits in Quito and taught at the Universidad de Lima. It is said that Magnin gave him a map he himself drafted of the Spanish missions of Maynas.[11] It might be that this exchange stimulated the Jesuit priest to write, in 1744, *Millet en harmonía con Descartes o Descartes reformado* (Millet in Harmony with Descartes, or Descartes Reformed), in which he uses Newtonian laws to defend the Copernican system. In his reports, La Condamine stresses the usefulness of the data Magnin gave him concerning the rubber tree and other species used by the natives as narcotics. Magnin shipped this work to Europe twice, in 1744 and in 1747, since the first shipment was lost in a shipwreck.

It could be said that Magnin, elevated to the rank of corresponding member of the Académie Royale des Sciences (Royal Academy of Sciences) of Paris, is one of the most successful examples, if not the first, of the aforementioned Jesuit adaptability to French Enlightenment thought in New Granada. Although we do not intend to analyze in detail this adaptability, we do want to point out that the internal atmosphere of change experienced at the universities of the Ignatian community in Quito, heightened by the external factor of the visit of the geodetic mission, was helpful. So, despite local difficulties, Magnin's activity gradually gave him access to theoretical viewpoints that were less eclectic and more committed to the new cause.

Juan Manuel de Velasco y Petroche, a Creole from Riobamba, was a contemporary of Magnin's. He was educated at the Colegio de Popayán and entered the Jesuit community in 1746. He stands out for the scientific recognition he achieved and for his controversial *Historia del Reino de Quito* (History of the Kingdom of Quito). His compatriot Pedro Vicente Maldonado was also an outstanding geographer and naturalist who studied in Quito with the Jesuits. A friend of La Condamine and Magnin, he joined (in 1743) the former on his return trip to Europe, "going up the Amazon and charting its course." In 1746, he was named a member of the Académie Royale des Sciences, and two years later he joined the Royal Society of London, but death surprised him before he could claim this position.

As we have noted, the geodetic mission stimulated the Quito elite's interest in observation and scientific experimentation. Starting with the activities of the mission, elite culture became more solidly modern, and the elite acquired a vision of its territory and of itself and a clearer idea about Quito's place in Latin American geography. The first technically skilled geographical charts were drafted during this time, among others, the *Breve descripción de la provincia de Quito* (Brief Description of the Province of Quito), published in 1740 by Father Magnin; the Jesuit Jacinto Butrón's *Compendio histórico de la provincia y puerto de Guayaquil* (Historical Summary of the Province and Port of Guayaquil), published in 1745; and especially the work of Pedro Vicente Maldonado, *La villa de San Pedro Riobamba* (The Town of San Pedro de Riobamba, published between 1743 and 1747), *Descripción de la provincia de Esmeraldas* (Description of the Province of Esmeraldas, published in 1744), and *Mapa de la provincia de Quito* (Map of the Province of Quito, published in 1750).

As we mention in the introduction, Copernicus's and Newton's theories circulated in their practical form in the boundary commissions. These expeditions, which had as an overall goal the fixing of borders and the establishment of inland navigation routes, took along a lot of scientific instruments and important astronomers. The boundary expeditions to Brazil in 1750 was organized by a group of astronomers. The Spanish expedition was organized by Francisco Requena, J. F. Aguirre, Juan Varela, Diego de Alvear, and José María Cabrera.

There are numerous primary sources available on this subject, but the work of one man—Félix de Azara—stands out. Azara was assisted by demarcation astronomer Antonio de Pires da Silva Ponte. Especially worth mentioning are "Memoria del estado rural del Río de la Plata" ("Report on the Rural Condition of Río de la Plata") and the extraordinary 1798 map of Brazil entitled "Nova Lusitania" (New Lusitania).

We should also point out the use of scientific instruments by people not linked to the aforementioned missions. Very interesting is the observation of Mercury's passage in front of the sun on November 5, 1789. We know that "Leverrier made use of the figures obtained by Alcalá Galiano and Vernacci at Montevideo for calculating Mercury's perihelion, the anomalous disposition of which would appear in the General Theory of Relativity."[12]

Antonio de Córdoba's expedition to the Strait of Magellan employed officers who had astronomical training based on Newtonian physics. This mission was carried out in two stages between 1785 and 1789; it completed the charting of the strait and an analysis of this route as compared to the sailing through Cape Horn via the Pacific route.

There were expeditions to explore the Colombian coast prior to Mutis's arrival. French clergyman Louis Fevillé, a mathematician and naturalist, arrived in Cartagena in 1704 and is considered the first naturalist in the Viceroyalty of New Granada. Dutchman José de Jacquin, a pupil of Antonio Jussieu's, was sent to America by the emperor of Austria and carried out important studies on the plants on the coast of Cartagena and the Caribbean between 1752 and 1759. Clergyman Juan de Santa Gertrudis Serra, another botanist, arrived in the Viceroyalty of New Granada between 1756 and 1765 and listed the plants of the Atlantic coast and the Magdalena valley with great accuracy and detail.

As happened in Quito, it is not by chance that in Santa Fe the teaching of Copernican theories first took place in the Universidad Javeriana in 1755, in a philosophy course entitled *Physica specialis et curiosa*.[13] It is evident that the Descartes-Newton debate was not absent from the universities of Santa Fe and Quito. In the writings that have reached us about these classes, mention is made of, for example, a scientific topic that was the source of controversy: the shape of the Earth. The heliocentric theses, that is, the law of universal gravitation that regulated all mechanical and natural phenomena, were also incorporated as the basis for Newton's world system. We can deduce that, starting from this theoretical concept, Newton formulated his prediction of the shape of the Earth.

Experiences like those of the missions to Peru and Lapland gave a factual base to theoretical predictions, strengthening the process of acceptance of Newtonian physics and the Copernican hypothesis. In the *Physica specialis et curiosa*, the debate on the shape of the Earth was reduced to the experimental results of the two expeditions; it is clear that the victory belonged to the English philosophical doctrine, in opposition to the French, or Cartesian, philosophy. The text presents the modern theory concerning the shape of the Earth.

This topic could not escape the consideration of Jesuits because of its direct link to members of the expedition. Besides, these priests should have been aware of the cultural adaptability of the order in Europe by virtue of their reading of the *Mémoires de Trévoux*, which circulated widely among intellectuals of the time.

Pressured by the social and intellectual impact of the geodetic mission on local Enlightenment circles, Jesuit teachers then dared to test the order's international educational project at the Universidad Gregoriana starting in the 1740s. As this project became viable in Quito, it was transmitted in the following decade to Santa Fe. The delay might be explained by Santa Fe's lack of a catalyst as powerful as the geodetic mission for new ideas.

Many circumstances had to converge in all universities. It is not clear that the Jesuits who received an old-fashioned education did not debate and incorporate into their pedagogical methods the adaptationist philosophy, even in those distant viceregal cities where the Peripatetic and systematic tradition prevailed (although in these centers the pressure for a hasty introduction of the new science was not as strong as in Europe). When we look at the implementation of the Jesuits' pedagogical methods, however, we should take into account that the Jesuits' institutional project for cultural modernization in their networks is one thing, its concrete application in the American viceroyalties by missionaries, quite another.

But what is indeed certain is that local changes and events cannot be understood outside of an international project, at least with regard to Jesuit changes in educational activity; each one was different and unique to its location. The Society of Jesus adapted its systematic traditions to Newtonian processes toward the late 1730s, when experience allowed the Jesuits to accept the new physics, at least in Europe, and no one continued to place theology and the new science in opposition. It was, rather, about defining their respective areas. Copernican theses and the system of the Newtonian world were mathematical constructions validated by experience and calculation, but only as hypotheses derived from a scientific way of explaining nature. Theology, however, reserved for itself the ultimate explanation of natural phenomena, not with hypotheses but with theses referring to the final cause: the creation and the regulation of the world by an omnipresent and omnipotent God.

The teachings of Jesuit Francisco Javier Aguilar at the Universidad Gregoriana provide another relevant example of how the new system was disseminated. In the philosophy course he gave from 1753 to 1756, Aguilar discussed general and special physics.[14] In his introductory section concerning "the world, sky, and meteors," he explained that special physics was a "pleasant treatise, by all means, which a philosopher cannot ignore without tarnishing his reputation."[15]

He taught natural phenomena in the context of the world and the sky. He also named what he considered the five main world systems—the Ptolemaic, the Copernican, the Tychonic, the Platonic, and the Egyptian—but discussed only the first three. He explained that the Ptolemaic system placed the immobile Earth at the center of the universe and that Copernicus "devised other systems of the world around the year 1477."

Father Aguilar admitted that the Copernican system seemed "appropriate for explaining the four variations of time, the year, and the stars' movement and setting." However, he pointed out that Copernicus's theses had been forbidden by the Inquisition since 1616, although one could "state them as hypotheses."

The third system he taught was that of Tycho Brahe, who opposed Copernicus's thesis and invented his own, "according to which the sun rotates around the Earth, as in Ptolemy's system, but Mercury, Venus, Mars, Jupiter, and Saturn rotate around the sun in epicycles." Aguilar was inclined toward Tycho's system, since "moderns place our Earth standing still at the center of the universe; according to them, the planets rotate in their own movement around the Earth, except two of them, Venus and Mercury, which, likewise with their own motion, rotate around the Sun, having it as their center."

Although viewpoints were primarily formed by a scientific spirit and according to local context and the impact of certain events in central countries, at times, not very well developed postures, such as that of Father Aguilar, acquired some importance in the Latin American environment. As is well known, the dominant theses, defended by the church, were the Ptolemaic, which followed the teachings of the holy scriptures literally. In Spain, the posture against the Copernican system was held by a majority around 1750, but an intermediate position concerning world systems—between tradition and modernity—appeared at this time, that of Tycho Brahe, and this position was adopted by some Jesuit professors at the Universidad Gregoriana.

As we have pointed out, in the Universidad Javeriana of Santa Fe, the professor of philosophy offered a course called *Physica specialis et curiosa* in 1755. Whoever he was, this Jesuit organized his teaching similarly to Father Aguilar's, but his arguments were different. Whereas Aguilar considered the system of Tycho the most appropriate, the professor of philosophy at the Universidad Javeriana pointed out in his "second argument: of the constitution of the main parts of the world" that the

Copernican system is disseminated openly today in Italy, France, and some regions of Germany.

This Jesuit followed the logical order of explaining the oldest systems first: the Pythagorean and the Platonic. He then set forth what he called "some new systems": the Tychonic and the Copernican. He refuted Tycho's way of calculating time (adopted by some Aristotelian scientists) as being "thoughtless" for not taking the Sun as a fixed system of reference. He stresses in the *Physica* text that "Copernicus's system is the simplest," for explaining the solar system, compared to previous models, in particular, Ptolemy's system of cycles and epicycles with the Earth at the center. Even simpler than the Ptolemaic system was that of Tycho Brahe. Thus, the efforts of Copernicans such as the author of the *Physica specialis* and Mutis in the eighteenth century were focused on refuting Tycho's system.

As we have mentioned, in the Jesuit universities of Quito and Santa Fe, the theories of the cosmos were freely discussed between 1740 and 1760. Father Juan Bautista Aguirre replaced Father Aguilar, the holder of the chair in philosophy. In the course he taught between 1756 and 1758, he defended his predecessor's theses. He explained Copernicus but preferred Tycho Brahe's theses "for not being against the holy scriptures." Father Aguirre is important for putting forward other scientific novelties supported by experience and modern bibliography: sunspots, comets, fire, the gravity and lightness of the elements, the states of water, the elasticity of air, the distance between the Earth and the Moon and between the Earth and other planets. In 1757, with his student José Linati, he publicly defended 257 philosophical theses. Thesis 16, dedicated to special physics, advances the idea that "Ptolemy's world system is contrary to what is observed in astronomy, and therefore it should be rejected. Copernicus's system is opposed to the holy scriptures, and, for the same reason, it should be refuted." Thus, according to Thesis 16, Tycho's system should be preferred to those of Ptolemy and Copernicus.

We should remember that, although Copernicus's *De revolutionibus* was indeed put on the *Index* (a list of books forbidden by the Inquisition) in 1616, around 1758, it was omitted from the revised *Index*. But cultural events of a normative order, even those as important as this one, were not necessarily immediately transferred to the social domain of institutions far from the center. Regardless of the local resonance of this fact, it is possible that some of the professors of philosophy in Quito and Santa Fe, at this stage of transition to new ideas and adaptation, chose, as did Father Aguirre, to continue defending Tycho's theories. In the intertwining of cognitive strategies characteristic of this period, and without the mediation of a preestablished Newtonian cultural field, different major figures headed alternative educational programs without necessarily opposing one another. While Fathers Aguilar and Aguirre promoted Tycho's theses, the Javerian Jesuit chose to defend the Copernican system.

José Celestino Mutis and Juan de Hospital Declare Themselves Copernicans

As we have pointed out, the formal liberalization of the teaching and dissemination of *De revolutionibus* did not guarantee acceptance by traditionalist groups of the heliocentric theories instead of those in which they had been trained. They preferred medieval and classical viewpoints, like those of Ptolemy and Tycho.

In this context, it is quite significant that Father Juan de Hospital taught the Copernican system and Newtonian mechanics in the philosophy course at Quito from 1759 to 1762 and refuted the Ptolemaic and Tychonic systems. On December 14, 1761, Manuel Carvajal, under the guidance of his teacher, Juan de Hospital, defended twelve theses. The eleventh questioned Ptolemy's and Tycho's world systems and recommended the Copernican system because it postulated the movement of the Earth and because it was "the closest to astronomical observations and physical laws."[16]

According to Ekkehard Keeding ("Las ciencias naturales"), Father Hospital tried to establish a scientific circle in order to disseminate the basic practices of modern physics. We do not have enough information to establish a direct relationship between Father Hospital and the Academia Pichinchense, created in 1766; however, we can assert, following Keeding, that the physics course, taught from 1760 to 1761, had profound repercussions on the cultural activity of Quito's Creoles, like Eugenio Espejo, Carvajal, and Joaquín Rodríguez. Rodríguez was the father of Miguel Antonio Rodríguez, who introduced the teaching of anatomy into the physics course and made the study of mathematics at the Universidad de Quito important.

Before the Jesuits were banished in 1767, José Celestino Mutis had already propounded the Copernican system at the Colegio San Bartolomé without any public repercussions. Moreover, he emphasized the Universidad Javeriana's tradition of teaching new sciences and natural philosophy: "If love of truth has stopped me more than was fair from manifesting my inclination toward the Copernican system, there will be reason enough to conclude with celebration the happy time when we see the rebirth of natural philosophy taking place in this kingdom."[17]

The dissemination of the heliocentric theses and the Newtonian world system set forth by Mutis between 1760 and 1770 was most consistent with the paradigm defended in the home country and also of greatest local impact, but this was not the first time this had happened. As we have stated, before the arrival of Spanish reformist ideas created by Bourbon Enlightenment policies, local appropriation of the new science (ascribed to the Aristotelian and systematic viewpoints) was continually advanced in New Granada during the decades 1740–1760.

When Mutis started the institutionalization of the discourse of the new philosophy between 1762 and 1767, he certainly did not do so in a cultural space devoid of ideas on this subject. During the previous twenty years, the intellectual elite of Quito and Santa Fe, and their spheres of influence, had recognized the existence of technical developments and new data in physics and science. A new theoretical-experimental way of thinking slowly overcame the old philosophy concerning the mysteries of the natural world and even differed from that of modern Cartesians. Such differences, however, were not viewed as being in opposition or conflict in all cases. These new techniques and this new knowledge were often recognized, in their operational aspect, as instruments with practical explanatory value, and they could function without ontological doubt under Aristotelian and Peripatetic views. What Mutis would do twenty years after the birth of this local tradition was show that the difference between Newton and Copernicus, and that between Aristotle and Descartes, was epistemological and philosophically irreconcilable.

It begins to become clear, then, that the new physics could be neither submerged nor adapted to the old metaphysics, since it was based on its own natural philosophy. Mutis had also shown in his scientific activity and disseminated through his teaching new and eclectic attitudes toward modern philosophy, particularly concerning the delicate matter of the relationship between physics and natural religion, and that between mathematics and experience. His attitudes are explained by both the transitions he made in his intellectual training and the ambivalent circumstances of the local environment in which he undertook to spread the new science.

In 1767, at the Colegio San Bartolomé, Mutis presented his "Reflexiones sobre el sistema tycónico" (Reflections on the Tychonic System), in which he defends Copernican ideas by using two propositions: (1) that the Earth indeed moves like the other planets do, but the Sun and the stars remain static, except for a unique movement by the Sun on its axis; and (2) that the Copernican system is by no means opposed to the holy scriptures.[18] In general terms, "Reflexiones" is a discourse that we could place toward the end of the period during which the new philosophy was adapted to traditional world views. It is a clear attempt to make the Copernican theses viable, starting from the traditional options represented mainly by Aristotelian metaphysics and, secondarily, by the Cartesian system. Mutis almost certainly prepared this discourse keeping in mind the difficulties that conservative institutions, in the hands of the Dominicans, erected before every local project of cultural or educational reform. In spite of his eminence and academic prestige as professor of the new philosophy, Mutis was aware of the institutional force of the Dominicans and was careful not to provoke a more direct confrontation between theses.

The indirect confrontation was settled in 1773 with a discourse that was less eclectic and more committed to Copernicus and Newton. Between 1767 and 1773, there was time for Mutis's views to mature and for him to produce a systematic work that used direct study of Newton's physical and mathematical work to conceptually appropriate these theses.[19] In 1773, Mutis openly came out as a Copernican and dedicated his conclusions, stated publicly at the Colegio del Rosario, to the viceroy's wife: "Being well educated with fine knowledge and a clear understanding that I could never have gotten in the darkness of the old philosophy, I openly confess that I am a Copernican."[20]

Mutis presented sixteen theses to support the heliocentric system. What stands out is that he dedicated at least eleven to proving that these theories were not forbidden and that, on the contrary, teaching them was ordered in the new reforms proposed by Charles III. Therefore, "Reflexiones sobre el sistema tycónico" is a text in which the cognitive strategy for the defense of Copernican thought articulates theoretical and conceptual arguments within normative and juridical rules and makes a clear case for the feasibility of Mutis's theses.

There was also a debate initiated in the City of Caracas between the aforementioned Jesuit professor at the university, the Count of San Javier, and a certain philosopher Valverde of noble condition and ecclesiastical state, in the king's active service and graduated from the Universidad Tomista.[21] The dispute began on August 1, 1770, after an argument between Valverde and the count about the usefulness and uselessness of Aristotle's philosophy. Valverde responded in writing on August 7, fixing his philosophical position. Valverde's argument proves that he was familiar with Copernican and Newtonian theses and with the modern discourse of new science. Nevertheless, we have no reason to believe that particular and already established culture was based on concrete knowledge.

Valverde's work contains a theological and critical case against the old philosophy from the new science's viewpoint, but the document lacks a sound foundation. It is clear that Valverde was familiar, as was common for a scholar in the colonial villages, with European works and cultural events, in his case, those from England. From some vantage points, we can see that his argument was based on the authority of someone like Robert Sanderson,[22] and he quotes the bulletin of the Royal Society of London. But, we insist, his fundamental premise is theological rather than scientific. Valverde tries to use what seems to be new science to introduce his theological position, for example, when he refers to the principle of transitivity of equality—if A = B and B = C, then A = C—to attack the heretics who refused to accept the Holy Trinity.[23] Valverde also states his position as an unconditional follower of Thomistic philosophy.

Mutis's case is different from Valverde's, at least as regards the former's 1773 dissertation. That work was outlined in scientific terms to respond to any inquisitors who would debate his theories, as indeed happened. It was not, however, a courtly, incidental debate for amusement, as was Valverde and the Count of San Javier's.

In the process of spreading modern scientific ideas about physics in the colonial age, we also find several Enlightenment authors whose works should be read between the lines. They may have hidden their true ideas for fear of the Inquisition, which local intellectuals viewed as quite powerful in Spain because of its actions and, primarily, for the victims it continued to claim. Pablo de Olavide's trial provides a good example. This case embodies "the problems of freedom and culture, and primarily it showed the enormous power of the Inquisition in Spain."[24]

Dominicans Concede in Quito and Attack in Santa Fe

The Dominicans' attitude regarding public conclusions about the Copernican system was totally different in Quito from in Santa Fe. The difference was marked by the correlation of political forces and, particularly, by the control of education.

We can establish a first stage in both Quito and Santa Fe, between 1740 and 1767. During this period, Mutis and the Jesuits freely propounded the Copernican system in public; the Dominicans answered with holy silence. Among possible causes there is, first, the power of the Jesuits within the Inquisition. Inquisitor General López de Prado, a friend of the Jesuits, occupied this position from 1746 to 1767. Interests and political-educational power were well defined at the time, however, and both orders respected one another. Indeed, both the Jesuits and the Dominicans had enjoyed the same rights and privileges to grant degrees in their respective universities since 1702.[25]

To understand the Dominicans' polemic against Mutis after he began to espouse Copernican theses, we need first to present some background. By the early 1770s, things had changed with the expulsion of the Jesuits. The Dominicans thought they could occupy the void and, for example, petitioned the Junta de Temporalidades for the goods and educational privileges of the "expelled" Jesuit community.

Quito's reality was different from that of Santa Fe regarding the creation of a public university. The Junta de Temporalidades undertook the founding of a new public university using the possessions of the Universidad Gregoriana and the Jesuits' Colegio-Seminario de San Luis. The Dominicans satisfied their long-standing ambition of controlling and monopolizing university education in Quito. On August 23, 1776, the Junta agreed to transfer the Universidad de Santo Tomás, with its rents and property, to the building of the Colegio-Seminario de San Luis, declaring it the only official university in the Audiencia of Quito. The Dominicans also got the royal warrant of 1786, which stated that the patron saint of the new public university would be Thomas Aquinas.

In Santa Fe in 1768, Dominicans argued against Moreno y Escandón's efforts to restructure the program of studies and to erect a public university in the City of Santa Fe. They believed that the new university would eliminate their education privileges in the viceroyalty. The fiscal was following orders to reform the universities and give the king, rather than the pope, control over education. By transferring control of education to the state, the laity acquired the right to scrutinize education and the teaching and uses of educational institutions.

Moreno stated, in the aforementioned plan, that "clergymen have seized command of the sciences, they dominate in the employment of university rectors, supervisors, degree examiners and referees, and subject laypeople to hard servitude, to always living as inferiors without hope of ever shaking off such a heavy yoke."[26] Here the main reason for the dispute becomes evident: the open fight between the civil sector, which wanted to control education, and the religious sector, in this case, the Dominicans, who would not tolerate the loss of the privileges they received from their ten-year exercise of a monopoly in education. Moreno's proposed reform of education held that "the order of preachers is the only one that, deprived of the privilege of conferring degrees, will perhaps be able to pretend indifference after it takes charge of the aforementioned powerful motives."[27]

In this context, it is understandable that the Dominican opposition to Mutis's defense of the Copernican system in the 1770s involved more than philosophical differences. Their antagonism involved the most important topics of the day: the creation of a public university, and the issue of secular control of this university. Therefore, within a seemingly superfluous debate in an overly ideological intellectual atmosphere was hidden an intense confrontation between two organized groups that wanted to control education.

By debating with Mutis, the Dominicans were trying to gain time to consolidate their influence in Madrid, with the purpose of paralyzing Moreno y Escandón's educational program, which severely attacked their interests. The debate on the Copernican system was immediately linked to the dispute over educational reform. This time, however, the disagreement passed through the Inquisition of Santa Fe, then the Santo Tribunal (Holy Tribunal) of Cartagena, and, finally, the Inquisición Suprema (Supreme Inquisition) of Castille, on March 6, 1775. Apparently, Mutis ended up convincing his judges that the heliocentric system was not contrary to dogma.

Creoles in Quito, Caracas, and Santa Fe Return to Newton and Copernicus in the 1790s

Although it is true that the Enlightenment elite in the court of Charles III promoted secularization of the universities in both the juridical and the institutional fields, implementation was actually different. On one hand, proponents of the Enlightenment showed interest in university reform, yet they were not always able to contain the attacks on these reforms by certain religious orders and conservative sectors. In addition, there were other factors in the complex situation in Spain, and religious and Inquisitorial power alone was quite strong. The political atmosphere was yet another factor after the French Revolution. During the 1790s, the Spanish Crown developed measures to control and repress the professors, students, and books that were circulating in the viceroyalty.

The situations in Quito, Caracas, and Santa Fe differed. The development of universities, particularly in these capitals, took separate paths after the expulsion of the Jesuits.

Santa Fe de Bogotá, capital of the viceroyalty, had twenty thousand inhabitants in 1774. According to the census carried out by Moreno y Escandón in that year, the city had four institutions of higher education: the Universidad de Santo Tomás, the Colegio Mayor del Rosario, the Colegio Mayor de San Bartolomé, and the Universidad de San Nicolás de Bari. The only institution that could grant degrees to laypeople after the expulsion of the Jesuits was Santo Tomás.

Quito, capital of the Audiencia, also had 20,000 inhabitants around the middle of the eighteenth century; by 1780, it had 28,451, but only one university—Santo Tomás—that the secular population could attend after the expulsion of the Jesuits. This university became public in 1775, and in 1786, it was renamed the Universidad Pública de Santo Tomás. The first attempts to secularize it started in 1788, upon the nomination of a lay rector and three professors for the philosophy chair. The power of the bishop's money was stronger, however, and once again the university came under religious control. At the beginning of the nineteenth century, the rector and the board began a new campaign for independence from the control of religious powers.

The Universidad de San Fulgencio in Quito had papal and royal support. It, however, conferred secular degrees only until 1775.

In Caracas, three convents were dedicated to teaching: San Francisco, San Jacinto, and Las Mercedes. Dominicans and Franciscans also opened their classrooms to laypersons. The only university that granted degrees during the colonial period in the City of Caracas was Santa Rosa de Lima (with the *colegio-seminario* to which it owed its academic origin). In 1784, the *colegio-seminario* separated from the university and took on a more secular character, although every two years the rectorship was shared with a clergyman.

The difference regarding institutions and monopolies was clear between Quito and Santa Fe, primarily. In Quito, Dominicans were the lords and masters after the expulsion of the Jesuits, while in Santa Fe, they continued to fight with the civil sector for the education monopoly. In Caracas, their influence in the university was not as strong because of its robust secular character, especially after 1784. They confronted the Franciscans and the civil sector to increase their perquisites in the university.

In Quito, Father Hospital's supporters began to stand out in the Audiencia around 1779. Francisco Javier Eugenio de Santa Cruz y Espejo, a pupil of Hospital's in the 1761–1762 year, graduated in 1767 in medicine from the Universidad Santo Tomás. In 1779, he published *El Nuevo Luciano de Quito* (The New Quito Lucian), in which he criticizes the reigning philosophy.

Although Espejo did not hold any professorship in the university, he guided many Creoles of the new generation as director of the library.

One of these was Miguel Antonio Rodríguez, son of one of his classmates in philosophy courses taught by Father Hospital. Rodríguez, as part of the generation that attempted to reconcile scientific and religious ideas, put Copernicus in the spotlight again in a philosophy course in which he taught Newton and Copernicus. After twenty-five years of not openly defending the theories of heliocentricity in Quito, in 1797, Rodríguez and a student, Pedro Quiñónez y Flores, published their conclusions as "Theses Philosophicas Sives Philosophia Universa" [Philosophical theses following the universal philosophy]. Rodríguez also introduced the teaching of anatomy and published a declaration of human rights in 1813.

It is somewhat surprising that the Dominican community did not speak out against Professor Rodríguez. We believe that this confirms our conviction that, beyond maintaining a degree of philosophical orthodoxy, what most interested the Dominicans was control of the universities in Quito and Santa Fe. The situation in Caracas was similar. As we shall see below, the Dominicans remained silent about Marrero's debate; we believe, however, that they would have been reluctant to teach the new Enlightenment current in their philosophy course at the Universidad de Caracas.[28]

In this regard, let us remember that, although the Universidad Santo Tomás of Quito became a public institution in 1776, the new Constitution of 1786 guaranteed the Dominicans that the patron saint of the institution would be Saint Thomas. Father Pérez Calama's changes in the course of study in 1791, but only partially enacted, ensured that the professorships in theology—Prima and Víspera—were held by a Dominican. This also occurred at the Universidad de Caracas, where Dominicans had controlled the professorships in philosophy and holy scriptures since 1742.

In the Theology Department under Pérez Calama's program of study, students studied "Saint Thomas himself." The department urged, "Let us form a system for allowing the boys be imbued with what the saint clearly teaches. Using this method, the students will become great and solid theologians."[29]

The Copernican system was not part of the study of the physics. Dominicans, possibly leaning on the Constitution of 1786 and on Pérez Calama's program of study (and before the drafting of new bylaws for the university in 1803), engaged in a decisive debate with the rector and professors in which they opposed eclecticism and argued for uniformity in teaching, centered on the doctrine of Saint Thomas. Apparently, by this time, the Dominicans' influence in the Universidad de Quito had declined, because the rector and the board supported eclecticism and the study of mathematics and physics.[30]

The educational atmosphere in Santa Fe around 1791 is demonstrated by, first, Francisco Antonio Zea's publication of "Avisos de Hebephilo" (Warnings of Hebephilo), a critique of scholastic teaching,[31] and, second, Colegio San Bartolomé students' request that a professor of mathematics and natural philosophy be appointed at their expense.[32] Furthermore, the *Papel Periódico de la Ciudad de Santa Fe de Bogotá* (Periodical Paper of the City of Santa Fe de Bogotá) began publication under the direction of a Cuban, Manuel del Socorro Rodríguez, director of the public library and coordinator of the Tertulia Eutropélica (Gathering of Good Taste). Other meetings include Nariño's Arcano de Filantropía (Mystery of Philanthropy), and Doña Manuela Santamaría de Manrique's Tertulia del Buen Gusto (Gathering of Good Taste). Also in 1791, clergyman José Domingo Duquesne wrote *Historia de un Congreso Filosófico tenido en Parnaso por lo tocante al imperio de Aristóteles* (History of a Philosophical Congress Held in Parnassus concerning the Empire of Aristotle). This document's confrontational style is similar to that of those that Espejo was publishing in the City of Quito. Thus, we can conclude that Enlightenment thought spread among the Creole elite by means of literary circles, newspapers, and documents, and in institutions such as the Expedición Botánica, but it was not usually accepted at institutions of higher education.

The debate between Prof. Manuel Santiago Vallecilla and Rector Santiago Gregorio Burgos of the Colegio del Rosario is important because of its repercussions in the university environment in Santa Fe. Professor Vallecilla defended the method of useful philosophy included in Moreno y Escandón's program and would not follow Burgos's orders because of what Vallecilla called the futility of the Colegio del Rosario's rector's arguments in controlling the work of the professors and in qualifying the doctrines that they are to read.

In 1796, Prof. Juan Francisco Vásquez Gallo took a similar position when he claimed that he "did not want to read or defend the *Summa Doctrina* by Saint Thomas and Professor Goudin" and, instead, defended the heliocentric system. As he had already done in the case of Professor Vallecilla, Viceroy Ezpeleta supported the rectors instead of the professors who defended Enlightenment ideas. On June 15, 1796, the Junta de Estudios (Curriculum Committee) condemned "Professor Vásquez's excesses and personally whosoever might dare in the future even present as a hypothesis a system as sacrilegious as Galileo's."[33]

Even though the Junta de Estudios censured Vásquez, he continued to defend the heliocentric system and attacked Goudin, the Peripatetic. The rector of the Colegio del Rosario suggested as the only appropriate solution "throwing him not only out of the school but also out of the city, so that he does not corrupt with such influences and such fatal disobedience."[34] Vásquez's tenacity in maintaining his position reveals the consciousness of the scholarly elite of Santa Fe about the unyielding cultural value of certain theories. They believed that these should be taught even, and above all, in official institutions, even if it meant challenging the power of what was perceived as the old order.

Of course, retaliatory measures did not eliminate the controversy about the Copernican system in Santa Fe. The debate resurfaced on June 20, 1801, when, at the request of Viceroy Mendinueta, Mutis prepared a report on Copernican theses that the Calced Augustinian Hermits defended in public at the Universidad de San Nicolás de Bari. This university had become known for teaching new theories after its reform by Peruvian priest Francisco Javier Vásquez in 1773. Vásquez had distributed pamphlets demanding the "banishing, the outright casting out from the schools of Peripatetic philosophy and theology, filled with useless and quarrelsome questions, good for nothing but wasting time, and the teaching of a useful and profitable philosophy, able to empower students for the fruitful study of other subjects, and to find the truth—the end toward which all aspire."[35]

The new method, which opposed Scholasticism, began to be taught in the Universidad de San Nicolás de Bari in June 1776, when university ecclesiastical studies were reorganized in the convents of the Calced Augustinian Hermits of Cartagena and Santa Fe. In the latter, Creole and Calced Augustinian Diego Francisco Padilla was elected head of the philosophy department, which position he held from 1776 to 1782. On returning from a trip to Europe in 1788, he was named rector of studies at this institution and reorganized the teaching of philosophy by introducing systematic instruction in the new ideas. In 1800, his efforts were secured with the establishment of the Mathematics Department and the teaching of Newton and Copernicus. It is therefore understandable that, in his report to the viceroy in 1801, Mutis sanctioned, in a totally modern discourse, the conflictual process whereby the teaching of Copernicus and Newton in Santa Fe would be legitimated.

Around 1790, the situation in Caracas was equally difficult for those who took on the promotion of Enlightenment ideas on university boards. There was an open confrontation between Enlightenment sectors and the Scholastic tradition supported by the Crown. This was a time of political repression in the American colonies, with the home country manifesting its decision to exercise total control over courses of study, professors, students, and textbooks. Humboldt points out that, in these years, "the setting up of printing presses was prohibited in cities of forty to fifty thousand inhabitants; peaceful citizens, who retired to the countryside and secretly read the works of Montesquieu, Robertson, or Rousseau, were suspected of harboring revolutionary ideas."[36]

The debate led by Prof. Cayetano Montenegro and a colleague, the Creole clergyman Baltasar de los Reyes Marrero, does not represent an isolated instance in Caracas or Spain. The seemingly insignificant expulsion of the son of Don Cayetano from the Mathematics Department for not doing his homework within the modern focus was enough for the Consejo Real to use him as an example. The regime had correctly noticed that, behind the teaching of new theories that aimed at practical applications, there lay the interests of the social-climbing elite and the subversion of the colonial order—both of which were impermissible. Thus we can easily understand the rigid (and draconian) measures used to control professors, such as the inspection of the class notebook, which students had to learn by heart before each session, and the bimonthly visits to the classrooms and the homes of students to "snatch" any book harmful to religion and the Crown.

These repressive policies, which were particularly heavy-handed in Caracas, also addressed the fact that, because of its geographical position and the free trade that developed there starting in 1788,[37] the city had become an entry point for books about the new philosophy and notable for their circulation.[38]

While royal decisions to repress professors in Santa Fe and Caracas shared traits, in Caracas, the members of the Audiencia and the rector supported Marrero, and the judgment and fine came from the Consejo Real. In Santa Fe, neither Vallecilla nor Vázquez Gallo was supported by the rector or the Junta de Temporalidades; they received support only from the vice-rector and, later, Camilo Torres, the precursor of independence. In Santa Fe, Viceroy Ezpeleta categorically and unwaveringly supported the rectors who represented traditional Scholastics. The Consejo Real took a similar position when it tried Marrero and supported Professor Cayetano and the Papal Chancery in spite of the backing that the Audiencia and its illustrious rector, Juan Agustín de la Torre, gave Marrero. It should be kept in mind that the town council, with the support of the university board, had been pushing for the creation of a mathematics department for several years. As in Santa Fe, Caracas's Creole elites discussed and put into practice, inside and outside university classrooms, the useful knowledge of the new philosophy, using their initiatives to fill all the spaces that could be useful to their social and cultural project.

By the time of this debate in 1790, the Academia de Derecho Público y Español (Academy of Public and Spanish Law) had been created in Caracas, with Miguel José Sanz as its first president. He was known in the history of Venezuela as a remarkable lawyer, founder of the Colegio de Abogados [Bar Association] of Caracas and the Academia de Derecho Público [Academy of Public Law], adviser of the Real Consulado, accurate critic of the stagnant colonial educational system, journalist for the *Seminario de Caracas,* and precursor of the entire political reform.

This same year, the rector of the Universidad de Caracas, Dr. Juan Agustín de la Torre, published and circulated the "Discurso económico, amor a las letras en relación con la agricultura y comercio" (Economic Discourse: Love of Literature as Related to Agriculture and Trade). In this paper, the rector outlined, as had Zea in Santa Fe, the urgent need to develop a "patriotic science." He also marked "science as an instrument linked to life, possessing the secret to strengthening and to dignifying it."[39] Mathematics was vital to the development of all sciences; therefore, de la Torre stressed, as Mutis had done in Santa Fe thirty years before, that it was necessary to create the department, "contributing the books, instruments, and machines indispensable to this teaching, because without this aid, it will be difficult to show the youth the effects of its application."[40]

For his part, Marrero taught, among other fundamental theses, that experimental philosophy should be preferred to the merely rational, because the first rests on reason and experience, while the last rests only on reason. This Enlightenment professor expressed several public conclusions defending the Copernican and Newtonian systems: Kepler, Huygens, Volta, Lavoisier, Musschenbrock, Buffon, Sigaud, Bails, Jacquier, Nollet, Brisson, among others, find that their theses are defended in the Universidad de Caracas, not only in the period when Marrero taught (1788–1790), but also presented and defended by Professor Pimentel and other teachers who continued teaching in the Philosophy Department.

In general, we can point out that, in Quito, with the teachings of Hospital, as well as in Caracas, with Marrero, and in Santa Fe, with Mutis, Félix Restrepo, and the application of Moreno y Escandón's plan, new theories were not lost in a vacuum. A new generation was trained and accepted experimental philosophy. Rafael Escalona, among Marrero's students, in fact continues to teach along the lines of his mentor's philosophy at the Universidad de Caracas.

Furthermore, while the new philosophy did not find institutional support in the universities, and was even repressed, we should point out that the scientific expeditions of the late eighteenth century, such as those of

Ruiz y Pabón to Peru (1777–1788), Martin de Sessé y Lacaste, Vicente Cervantes and the Creole José Mariano Mociño to Mexico (1787–1803), and Alexander von Humboldt and Aimé Bonpland, contributed, through scientific exchange, to the development of the natural sciences and the creation of societies and institutions that furthered the new theories.

The founding of the Reales Jardines Botánicos of Santa Fe (1783) and Mexico City (1788), the botany and chemistry courses taught in the Real Jardín Botánico of Mexico, the Chemistry Department in the Real Seminario de Minería of Mexico (1786), the Gabinete de Historial Natural (Office of Natural History) of Guatemala (1796), and the anatomical amphitheater of the San Andrés de Lima hospital (1792) also had significant effects, as did the newspapers that the organizers and members of these societies founded in which to express their thoughts and to spread news of their scientific advances.

Literary gatherings were ever present as well: that of the Uztáriz family in Caracas, the Concordia (concord) in Quito, and the Buen Gusto y Arcano de la Filantropía (Good Taste and Arcanum of Philanthropy) in Santa Fe. They also reflected the scientific and political curiosity of the Creole elite of the late eighteenth century.

Finally, we have to mention the role played by the Sociedades Económicas de Amigos del País not only in the economy but also in the dissemination of Enlightenment thought. They allowed "the 'Enlightenment' group, with the characteristics that defined it, to occupy the positions of power. They also incorporated their ideology and a new concept for social and economic relations and planned economic policy consistent with the ideals and purposes of Enlightenment despotism."[41] The first of these societies in Spain was the Vascongada, created in 1765. In colonial Spanish America the first was the Mompox, furthered by Mutis in 1785, its first director being Gonzalo José de Hoyos; it was followed by the Sociedad de Lima, created in 1787, the Sociedad de Quito in 1791, the Sociedad de la Habana in 1792, the Sociedad de Guatemala in 1795, and the Sociedad de Bogotá in 1801.[42]

Notes

1. Restrepo, *Historia*, p. 28.
2. Tovar Zambrano, "El pensamiento historiador," p. 35.
3. Archivo General de Indias (hereafter AGI), "Expediente sobre la Universidad Pública de Quito," Consejo 16 de abril, 1800, Sala Segunda, f. 13 (all folios and documents numbered by the authors unless otherwise noted). The Royal

Order of April 14, 1786, allowed the naming of alternate rectors, secular and ecclesiastic, at the Universidad de Quito. The Royal Edict of October 7, 1784, had already established a norm for the University of Caracas. The bishop offered to support the professorships if they were moved to Quito; thus, ecclesiastic economic power subjugated the civil sector.

4. A visit to the Universidad de Caracas was organized under the Royal Order of May 4, 1815, in order to reform constitutions and design a new curriculum. This plan and visit were assigned to Don José Manuel Oropeza, lieutenant governor of the Audiencia of Caracas. On December 20, 1815, the lieutenant presented the new program (AGI, Audiencia de Caracas, Sec. V, leg. 109, doc. 11).

5. González, "El rechazo," pp. 94–114.

6. "Expediente sobre la Universidad Pública Santa Fe, 30 de junio de 1777," AGI, Audiencia de Santa Fe, Sec. V, leg. 759, doc. 13, f. 15.

7. "Informe del fiscal Moreno, 25 de octubre, Santa Fe, 1771," AGI, Audiencia de Santa Fe, Sec. V, leg. 759, f. 11.

8. Minguet, *Alejandro de Humboldt*, p. 317.

9. The Jesuits established the Universidad Javeriana in Santa Fe in 1621 and the Universidad de San Gregorio in Quito in 1622. These universities conferred degrees on laypeople. In Caracas, the Jesuits did not teach at the university.

10. Vargas, *Polémica universitaria*, p. 11.

11. Lafuente and Estrella, "Scientific Enterprise," pp. 20–28.

12. Sala Catalá, "La ciencia," pp. 1584–1585.

13. *Nueva filosofía natural*, p. 52. Pedro Nel Ramírez believes that the author of this manuscript was a Spanish Jesuit. We know from Stella Restrepo that, in the eighteenth century, Mesuand, a Jesuit pupil of Descartes', was let go from the professorship he held at the Colegio San Bartolomé for teaching the new theories. We have been unable to establish a direct relationship between the author of the manuscript and Mesuand.

14. From the middle of the seventeenth century, and especially during the eighteenth century, the distinction between general and special physics followed the general outlines of Aristotelian classification. General physics included the fundamental subjects of the *Physica*, e.g., the essence of the natural being, the causes, the infinite. Cartesian treatises included, among other subjects, extension, quantity, and location. Special physics treated the heavens, generation and corruption, meteors, and even the soul.

15. Negrín and Soto, "El debate," pp. 50–51. It should be pointed out that Aguilar reproduced the generally accepted version of Tycho's system after the prohibition of Copernicus's system by the church in the early seventeenth century. All quotations from Aguilar in the next several paragraphs appear on these pages.

16. Carvajal, "Lo que se debe probar," defended during the term of Father Juan de Hospital, Quito, December 14, 1761, with the approval of Carvajal's superiors.

17. Hernández de Alba, *Pensamiento científico*, p. 110.

18. Mutis, "Documento," p. 65.

19. Arboleda, *Newton*; idem, "Acerca del problema."

20. "Alocución de Mutis sobre el sistema copernicano en el Colegio del Rosario," AJB, Sec. Mutis, Santa Fe, December 1777, leg. 25.

21. See Caracciolo Parra, *Filosofía universitario venezolana*. There are not many bibliographic references on Valverde's activities. We do not know, for instance, whether he was Creole or Spanish, nor do we know what kind of appointment he had in Caracas. See also "Carta de Valverde al Conde de San Javier sobre filosofía," Aug. 7, 1770, Archivo del General Miranda, sec. Diversos.

22. Sanderson was a spokesman for Newton's theories at Cambridge.

23. Thus Valverde positioned himself in opposition to those English and European Newtonians who defended the unitary, heretic theses made by the author of the *Principia*.

24. Álvarez de Morales, *Inquisición*, pp. 130–131.

25. Jesuits and Dominicans kept up a long argument over academic prominence and the right to grant degrees in the seventeenth century. This debate went through critical stages, such as when the Royal Decree of March 2, 1655, established that neither order would have a university or confer degrees. This conflict was resolved on May 27, 1702, when the Consejo Real decided to grant equal rights to both religious communities; the pope accepted that decision on August 13.

26. "Primer informe del fiscal protector de Indias sobre el establecimiento de universidad pública en Santa Fe, 9 de mayo de 1768," AGI, Audiencia de Santa Fe, Sec. V, leg. 75, doc. 8, f. 3.

27. Ibid., f. 10.

28. Pérez Calama, "Plan de estudios," pp. 177–220.

29. Pérez Calama, "Plan de estudios."

30. Lanning, "La oposición," pp. 224–241.

31. Arboleda, "La ciencia," pp. 193–225; Saldaña, "Nacionalismo," pp. 115–129.

32. "Los estudiantes de filosofía del Colegio San Bartolomé solicitan poner a sus expensas un profesor de filosofía que los instruya en física, matemáticas, botánica e historia natural" (AHNC, Sec. Colonia, Fondo Milicias y Marina, Santa Fe, 1791, vol. 128, f. 200–201).

33. Hernández de Alba, *Crónica*, vol. 2, p. 301.

34. Ibid., p. 303.

35. "Circular de fray Bautista González, reformador de la orden agustiniana en la provincia de la Gracia, sobre la reforma de estudios del padre Vásquez la que debía aplicarse en la Universidad de San Nicolás de Bari" (AHNC, Sec. Colonia, Fondo Conventos, Santa Fe, 18 de octubre de 1773, vol. 47, f. 92v).

36. Minguet, *Alejandro de Humboldt*, p. 281.

37. Caracciolo Parra León, *Filosofía*, p. 57; "Expediente de la Real y Pontificia Universidad de la ciudad de Santiago de León de Caracas, capital de provincias de Venezuela, practicada en virtud de Real Orden del 4 de mayo de 1815" ("Expediente de visita Universidad de Caracas," AGI, Audiencia de Caracas, Sec. V, leg. 446, doc. 10).

38. Important libraries are listed in several wills filed in Caracas, with works by a great number of European Enlightenment thinkers. See Leal, *Nuevas crónicas*, pp. 453–481.

39. de la Torre, "Discurso económico," p. 227.

40. Ibid., p. 229.

41. Negrín Fajardo, *Educación popular*, p. 39.

42. Minguet, *Alejandro de Humboldt*, p. 323; Novoa, *Las sociedades económicas*, p. 139; Barras de Aragón, *Las sociedades económicas*, vol. 12, pp. 417–447.

Bibliography

Archives

Archivo del General Miranda. Caracas.
Archivo General de Indias (AGI). "Expediente sobre la Universidad Pública de Quito." Consejo 16 de abril, 1800. Sala Segunda.
Archivo General de Indias (AGI). Sección V. Audiencia de Caracas. Seville.
Archivo General de Indias (AGI). Sección V, Audiencia de Santa Fe. Seville.
Archivo del Jardín Botánico (AJB). Sección Mutis. Santa Fe. Madrid.
Archivo Histórico Nacional de Colombia (AHNC). Sección Colonia. Fondo Conventos, Santa Fe. Bogotá.
Archivo Histórico Nacional de Colombia (AHNC). Sección Colonia, Fondo Milicias y Marina, Santa Fe. Bogotá.

Secondary Sources

Álvarez de Morales, A. *Inquisición e Ilustración, 1700–1834*. Madrid: Fundación Universitaria Española, 1982.
Arboleda, L. C. "Acerca del problema de la difusión científica en la periferia: El caso de la física newtoniana en la Nueva Granada." *Quipu, Revista Latinoamericana de Historia de las Ciencias y la Tecnología* 4, no. 1 (1987): 7–30.
———. "La ciencia y el ideal de ascenso social de los criollos en el Virreinato de Nueva Granada." In *Ciencia, técnica y estado en la España ilustrada*, J. Fernández and I. González (eds.). Madrid: Ministerio de Educación y Ciencia, 1990.
———. *Newton en la Nueva Granada: Elementos inéditos sobre los orígenes de nuestra cultura científica*. Cali: Universidad del Valle, 1990.
Barras de Aragón. *Las sociedades económicas en Indias*. Vol. 12. Seville: Anuario de Estudios Americanos, 1955.
Caracciolo Parra, León. *Filosofía universitario venezolana, 1788–1821*. Caracas, 1989.
Carvajal, M. "Lo que se debe probar: El sistema de Copérnico como el más acorde con las observaciones astronómicas y las leyes de la física." Quito, 1761.
de la Torre, J. A. "Discurso económico: Amor a las letras en relación a la agricultura y el comercio." In *Nuevas Crónicas de historia de Venezuela*, I. Leal (ed.). Caracas: Biblioteca de la Academia Nacional de Historia, 1985.
González, E. "El rechazo de la Universidad de México a las reformas ilustradas (1763–1777)." *Estudios Revistas de Historia Social y Económica de América*, no. 7 (1991): 94–114.
Hernández de Alba, G. *Crónica de muy ilustre Colegio Mayor de Nuestra Señora del Rosario en Santa Fe de Bogotá, Centro*. Bogotá, 1938.
———. *Pensamiento científico y filosófico de José Celestino Mutis*. Bogotá: Fondo Cultural Cafetero, 1982.

Keeding, E. "Las ciencias naturales de la antigua Audiencia de Quito: El sistema copernicano y las leyes newtonianas." *Boletín de la Academia de Historia* (Universidad Central), no. 122 (July–December 1973): 43–67.

Lafuente, A., and E. Estrella. "Scientific Enterprise, Academic Adventure and Drawing-room Culture in the Geodesic Mission to Quito." In *Cross Cultural Diffusion of Science in Latin America*, J. J. Saldaña (ed.), pp. 13–31. SLHCT, Series Cuadernos de Quipu no. 2. Mexico City, 1987.

Lanning, J. T. "La oposición a la Ilustración en Quito." *Revista Bimestre Cubana* 53, no. 3 (May–June 1944): 224–241.

Leal, I. *Nuevas crónicas de historia de Venezuela*. Caracas: Biblioteca de la Academia Nacional de Historia, 1985.

Minguet, C. *Alejandro de Humboldt: Historiador y geógrafo de la América española, 1799–1804*. Vol. 1. Mexico City: Universidad Nacional Autónoma de México, 1985.

Mutis, J. C. "Documento sobre el sistema copernicano: Alocución en el Colegio de San Bartolomé antes de 1767." *Revista Correo de los Andes* (September–October 1981): 97.

Negrín, O., and D. Soto. "El debate sobre el sistema copernicano en Nueva Granada durante el siglo XVIII." *Revista Colombiana de Educación*, no. 16 (1985): 37–54.

Negrín Fajardo, O. *Educación popular en la España de la segunda mitad del siglo XVIII*. Madrid: UNED, 1987.

Novoa, E. *Las sociedades económicas de amigos del país: Su influencia en la emancipación colonial americana*. Madrid: Prensa Española, 1955.

Nueva filosofía natural: Physica specialis et curiosa. P. Nel Ramírez (transcription, translation, prologue). Bogotá: Biblioteca Colombiana de Filosofía, 1988.

Parra León, C. *Filosofía universitaria venezolana, 1788–1821*. Caracas: Universidad Central de Venezuela, 1989.

Pérez Calama, J. "Plan de estudios de la Universidad de Santo Tomás de Quito. 29 de septiembre de 1791." In *Pensamiento universitario ecuatoriano*. Quito: Banco Central del Ecuador, n.d.

Restrepo, J. M. *Historia de la Revolución de Colombia*.

Sala Catalá, J. "La ciencia en las expediciones de límites hispano-portuguesas: Su proyección internacional." In *Actas del V Congreso de la Sociedad Española de Historia de las Ciencias y de las Técnicas*, M. Valera and C. López Fernández (eds.). Vol. 3. Murcia: Promociones y Publicaciones Universitarias, 1991.

Saldaña, J. J. "Nacionalismo y ciencia ilustrada en América." In *Ciencia, técnica y estado en la España ilustrada*. Madrid: Ministerio de Educación y Ciencia, 1990.

Vargas, J. M. *Polémica universitaria en Quito colonial*. Quito: Pontificia Universidad Católica del Ecuador–Banco Central de Ecuador, 1983.

CHAPTER 4

Scientific Traditions and Enlightenment Expeditions in Eighteenth-century Hispanic America

ANTONIO LAFUENTE AND LEONCIO LÓPEZ-OCÓN

During the eighteenth century, a great number of European expeditions traveled all over Hispanic America.[1] The numerous studies of the subject during the 1990s have considerably improved our knowledge of the Enlightenment and of the role science and technology played in the economic and social development of the Spanish colonies, although there are signs of exhaustion among researchers. It has become trite to reiterate the geostrategic dimension, neomercantilist profits, or the tendency toward emancipation that marked eighteenth-century scientific expeditions. A new effort is needed to find new points of view within studies of European expansion. It is therefore worth remembering that there were bureaucratic organizations besides European crowns that were capable of substantial and sustained reconnaissance efforts in the territories.

The most recent literature focuses on expeditions such as Cook's, Bougainville's, Malaspina's, or Humboldt's and seems to have forgotten about two other expeditionary traditions. Thus, we cannot understand the work of Sánchez Labrador and Martínez Compañón or the work in New Spain of engineers like Constanzó and sailors like Ulloa. This trend has been reinforced by classifying all the exploratory projects carried out in the eighteenth century into a single category. This seems to us to diminish the varied motives and actions of those involved in the study and exploration of Hispanic America.

Our thesis is that ecclesiastical and viceregal administrations were not mere appendixes of the home country's power but were organizations with enough political and economic autonomy to promote their own expeditionary efforts according to specific objectives and cultural projects—which might have been different from those of the Spanish Crown. In general, most published materials are marked by an emphasis on the ratification of Hispanic culture when compared to so-called European and North Atlantic customs. Naturally, this diverts attention to the arrival of European scholars and seems to enliven the pessimism that has characterized studies of Latin American culture. Although scholars do not ignore the presence of significant pockets of modernity within the Spanish colonial system by the end of the eighteenth century, the general trend is to rue the low levels of institutionalization, the scantiness of agents of change, and the poor viceregal or home country and to capitalize on the endless studies of the territory and its mineral and botanical wealth.

These symptoms of exhaustion are also evident in researchers' inability to question the universality-of-the-Enlightenment thesis, which is implicit in most of these studies. Many facts that are difficult to account for are overlooked, for instance, in Mexico, Hidalgo's uprising against the Crown and his shouts of "Viva la Virgen de Guadalupe" (Long live the Virgin of Guadalupe), or when Unánue stopped speaking until San Martín came to Lima's gates. And how do we explain the high percentage of reformers dressed in cassocks in Hispanic America if we continue to insist on the contradiction between the Enlightenment and religion? This contemporary ratification of Hispanic America (and Spain) by using inventories of books in private libraries or texts with references to European authors always faces an insurmountable obstacle: the number of renowned scholars, innovative experiments, and deeds crucial for the advancement of science will always be as small as the effort is great to find them.

In order to overcome this situation, it might be useful to place the study of expeditions and their cultural impact on colonial Hispanic America within the notions of the Catholic Enlightenment and socioprofessional roles. By using these frameworks, we shall posit the existence of a differentiated ideological context that we consider crucial to the function of the three administrations fighting over income and work within the context of the orientation of scientific practice. Using these two pillars as a basis—plus one that we will not consider here, cultural regionalization—we shall discuss what was untranslatable and international in Hispanic America with regard to the process of globalizing modern science. This process explains not only the mechanisms of the international transmission of ideas and institutions but also those of idiosyncratic recreation and habituation to the local climate. Seen from the receptor center, the preexisting cultural substratum was fertilized (and distorted) by the different and foreign elements, generating a tradition that started a dialogue with the new but without necessarily perceiving change as a struggle between tradition and modernity, for the former would be a condition for the latter.

One of the main problems posed by the studies of the globalization of science is the identification of the components of the aforementioned cultural substratum. Hence our objective is to demonstrate (or to remind readers of) the existence in Hispanic America of three scientific traditions that, throughout the eighteenth century, generated different expeditionary strategies for exploring and arranging the vast colonial territories. These are related to the three main structures of political, administrative, and economic power during the colony: the clergy, the home country, and the viceregal government.

The Expeditionary Tradition of the Church

All through the colonial period, the influence of the church on Spanish America was extraordinary. Its actions were not limited to missionary tasks; it pervaded all economic, judicial, educational, and health-related organizations. As the key structure of the colonial system, it was historically able to administer vast territories and to manage an immense economic and financial fortune. It therefore enjoyed broad autonomy.[2]

Such vast power demanded a permanent interest in incorporating the language of modern economics, politics, and science into the characteristic conceptual framework of Catholic intellectuality. The church, aware of its material power and always a jealous defender of its prerogatives and privileges, did not spare any effort to control the educational system and create an intelligentsia that adhered closely to its interests.[3]

Since many religious acted, from the moment they arrived in the colonies, from messianic—and sometimes millennial—conviction, a belief in the uniqueness of Hispanic America and its inhabitants did not take long to spread. Those first missionaries had many reasons to be fascinated with Hispanic American nature and excited by the magnitude of the colonization enterprise. It is therefore not strange that recording the continent's wonders and claiming for it the privilege of being unique thrived. It is within this tradition that the continent's roots were enhanced and the main ingredients for a Creole identity slowly matured, as David Brading notes: "The American clergy acted as the moral and intellectual leaders of colonial society."[4] This was a culture of religious, patriotic, idiosyncratic, eclectic characteristics and therefore original, critical, and modern.

At the end of the seventeenth century, the clerical elite of the Hispanic world and of other countries involved in the Counter-Reformation started to incorporate new knowledge that allowed the development of the Catholic Enlightenment. This was not merely a cosmetic operation, or *aggiornamento*, as evidenced by the violent debates aroused by Jansenism and Cartesianism; it was the revival of, among many intellectual traditions, political Suarism, Erasmist humanism, Jesuitical pragmatism, and Cartesian Malebranchism.[5] No work expresses better the difficulties of this transition than one published by the Benedictine Benito Jerónimo Feijoo. His arguments for the experimental method and against accepted religious superstitions gained him great popularity and remarkable authority among Hispanic intellectuals who favored change. Nothing demonstrates this better than the wide distribution of his writings; before his death, between four hundred thousand and five hundred thousand copies of his works had been sold, a spectacular number even today.[6]

A particular vision of the nature and reality of the economic empire built by the church in Hispanic America demanded a permanent research endeavor that comprised numerous works by members of diverse religious orders and by the secular clergy: botanical, zoological, astronomical, cartographic, philological, and historiographic studies. Undoubtedly, the Jesuits were most remarkable,[7] but the scientific program was not the work of a single order; it was shared by all orders that performed missionary activities on the frontiers of the empire. Among their less-controversial results is the cartography of the hydrographic basins of the Río de la Plata and the Paraguay River, of the Orinoco and the Amazon, and of Patagonia, Araucania, and the Californias, that is, the northern and southern limits of Spanish America. These cartographers included not only personal observations but also those of missionaries who preceded them, thereby showing a capability for capitalizing on experience accumulated over time.[8]

In few regions was the exploratory labor of the religious as decisive as in Paraguay. The example of José Sánchez Labrador contains many of the elements that can be found throughout Hispanic America.[9] A member of the Society of Jesus, he dedicated more than thirty years to exploring the Río de la Plata and reopened direct communication between Paraguay and Potosí through Santa Cruz de la Sierra, with the double purpose of finding an alternative to the route that followed the Chaco River (through Tucumán) and breaking the isolation of the missionary province of Chiquitos (by using the Paraguay River). He left a record of these endeavors in a journal, *Diario o relación fragmentaria de los viajes desde la*

Reducción de Nuestra Señora de Belén hasta las misiones de los Chiquitos (Diary or Fragmentary Account of the Voyages from the *Reducción* of Nuestra Señora de Belén to the Chiquitos Missions, 1766–1767), which includes a map of the Paraguay River and its natural waterways between latitudes 15° and 25°. Later, in Italy, he completed the monumental *Enciclopedia rioplatense* (Encyclopedia of the Río de la Plata), divided into three parts: *Paraguay natural* (Natural Paraguay, six volumes of natural history), *Paraguay cultivado* (Agricultural Paraguay, four volumes on agronomy), and *Paraguay católico* (Catholic Paraguay, one volume on human geography).

The wide knowledge of nature and ethnography the Jesuits acquired was organized according to a plan that combined the three realms of nature and the culture of the human societies that inhabited each region in America. These "natural" and "moral" histories, written according to a pattern that had been used since the seventeenth century, placed all created beings on a vital continuum in which every piece had a meaning and interacted with all of the rest (we might say they were in communion). The Jesuits' vision was encyclopedic and aimed at a historiography of totality; therefore, all territories in which they were active, regardless of the magnitude of their resources or their inhabitants' talent, were deserving of such a work, of a literature that would integrate them into history, hence the order's interest in building a heroic native tradition through literature and linking it to the great chain of being (as shown by the Jesuits' first astronomical and naturalist studies). As a whole, they created a genre we can still follow through the writings of authors such as Alzate, Gamarra, León y Gama, Caldas, Espejo, and Flores. This was the implicit objective of *Orinoco ilustrado y defendido* (Enlightened and Defended Orinoco, 1741), a book in which Valencian José Gumilla displays the natural, civil, and geographic history of the river and its waterways and gives an account of the government and of customs of the river natives. The book provides novel and useful news about the animals, trees, fruits, oils, resins, and medicinal herbs and roots of the area's flora and fauna. *Orinoco ilustrado* also includes in the first part a general treatise on the geography, natural history, and ethnography of those Amazonian territories. Around the same time, Italian Pablo Maroni wrote the *Noticias auténticas del famoso Río Marañón* (Authentic News of the Famous Marañón River), an account of the hundred years of Jesuit activity in the Province of Quito, in the upper and mid-Amazon, from their posts in Borja and the upper Napo.

As previously stated, this intellectual effort was not limited to the Jesuits; French Dominican Jean-Baptiste Labat, after living for twelve years in Guadeloupe and other parts of the West Indies, published a similar study entitled *Nouveau voyage aux Isles de l'Amérique* (New Voyage to the Islands of America, Paris, 1722).[10]

These three cases illustrate the international and pluricultural nature of the expeditionary program promoted by the church, which, in turn, explains, for instance, the early arrival of Cartesianism to the Kingdom of Quito. The presence of the Italian Jean Magnin there is a significant yet modest example, if we consider the work done in Italy by Jesuits after they were expelled from the Americas. At this point, the cumulative nature of the scientific and humanistic work these professors, missionaries, and expeditionaries carried out must be stressed.[11] Filipo Salvatore Gilij, for instance, corrected and completed Gumilla's text in his *Saggio di storia americana* (Rome, 1780–1784). His pro-Spain position in the debates about the New World was the exception among the expelled Jesuits,[12] for this argument was his opportunity to project his Creole nostalgia onto a vast historiographic program that allowed the order to claim the lushness of American nature, to praise the past grandeur of their distant homeland, and to demonstrate the existence of an erudite tradition there. This would be the case of Ecuadorian Juan de Velasco, Chilean Juan Ignacio Molina, and Mexican Francisco Clavigero.[13]

There were areas of Hispanic America in which other religious orders stood out, for instance, the Franciscans of Santa Rosa de Ocopa, who explored the Amazon.[14] Multiple trips by Antonio Abad, José Amich, Manuel Sobreviela, and Narcís Girbal helped warn Peru's viceroy about Portuguese incursions and explain their contacts with Mainas Province and its governor, Francisco Requena, but such trips were also used by Peru's explorers—Hipólito Ruiz and José Pavón—and would be published in *Mercurio Peruano*.

Exploration was not carried out exclusively by the Jesuits, the regular clergy, or in border areas. Frequently, pastoral visits were used as an opportunity for geographic explorations, to look for resources to allow the development of the diocese. Such is the case of the chorographic description of plants and animals in the Bishopric of Trujillo in the nine volumes of *Colección de planos, estados y estampas, relativos a la historia general de este obispado* (Collection of Plans, States, and Aspects Regarding the General History of This Diocese), written by Bishop Martínez Compañón between 1782 and 1785 and including varied information on the geography, social history, folklore, natural history, and archaeology of northern Peru.[15]

The Viceregal Expeditionary Tradition

From the early days in Hispanic America, there was a remarkable concentration of technicians solving the challenges posed by urbanization, defense, communications, and exploitation of the colonies. From Mexico City to Lima—the first metropolises in the New World, joined by Bogotá, Havana, and Buenos Aires in the eighteenth century—the viceroys organized a broad plan of activity intended to ensure the control of and communication with the territories under their jurisdiction. The viceregal bureaucracy thereby attained remarkable autonomy.[16] On the whole, the technician groups at the service of the viceroys were efficient support for two different but complementary tasks. First, they served as counselors and directors of the projects related to the improvement of urban infrastructure and sanitation, cartographic exploration, the opening and working of mines, the design and construction of fortifications, the definition of law pertaining to metals or the minting of money. Second was activity for controlling and creating public opinion, supporting educational reform, and engaging in publishing by collaborating with the *Mercurio* in Lima, the *Papel Periódico* (Newspaper) of Santa Fe de Bogotá, or the *Telégrafo Mercantil* (Mercantile Telegraph) of Buenos Aires.

Although there was a trend to make professional positions stable, technicians did not have quite the status of specialized viceregal functionaries. Yet they moved in the circles of influence and power in the palace and they were co-opted for missions and other tasks. Aware of their own importance, they tended to adopt a utilitarian ideology as a mark of identity, thereby creating a capital-based technocracy that gained wide autonomy in decision making. The specialized nature of their knowledge contributed to their autonomy, as did their extraordinary mobility and professional versatility.[17]

Undoubtedly, their social prestige was increasing, as shown by the numerous appointments some of them received. Ambrosio O'Higgins, for instance, was captain general of Chile before he became Peru's viceroy. Miguel del Pino was captain general of Río de la Plata. Miguel del Corral was governor and intendant of Veracruz, and Agustín Crame was governor of Havana.[18]

Although the literature on these technicians is not very abundant and is quite scattered, we can state that the exploration of the territories was entrusted to engineers and sailors. Their activity grew substantially beginning in 1768, at the beginning of the political and administrative reorganization of the colonial system to increase the security of the Atlantic front after the conflicts with the British during the Seven Years' War (1754–1763). Thus, between 1768 and 1800, two hundred Spanish military engineers were sent to the colonies to explore and fortify the empire's vulnerable areas, draw maps and write reports about the geography (both of which made them fierce guardians of the cartographic patrimony),[19] and construct civil engineering works for the territorial coordination of the empire, such as bridges, roads, canals, lighthouses, factories, jails, warehouses, courthouses, hospitals, mints, bullrings, and churches.

The sailors in Hispanic America's fleets also served the viceroys. Within the framework of policies designed to maintain territorial integrity, numerous hydrographic expeditions were organized, some of which were directly promoted by the viceroys.[20] The works of Ens. José de Evia stand out in New Spain. He worked between 1783 and 1786 in Mexico and produced remarkable charts of the territory between Florida and Veracruz. As a reward for his efforts, he was appointed captain of the New Orleans harbor and commander of the Louisiana Guard. His labor was completed by Lt. José Antonio del Río, who, in 1787, traveled over the east coast of Florida in order to find a location for a tar and pitch enterprise to supply the Havana arsenal.[21]

Several explorations to the North Pacific were organized from the northwest coast during the second half of the eighteenth century in order to contain Russian expansion. Several expeditions that included naturalists were launched to encourage the colonization of the Californias, for example, in 1792.[22] Mexican physician and botanist José Mariano Mociño, José Maldonado, and draftsman Anastasio Echeverría were commissioned, on the viceroy's order, to join Adm. Juan Francisco Bodega y Quadra on an expedition to Nootka Island to study the animal, mineral, and plant wealth of the area and anything that could facilitate their commercialization.[23]

Francisco de Viedma explored the Patagonian coastline in the Viceroyalty of Río de la Plata, traveled the Chico and Chalia rivers, and explored the lake that today is named after him.[24] In Peru, Viceroy Manuel Amat organized four journeys to the South Pacific in order to limit French and English expansion. The south of Chile was explored several times by expeditions sent from Chiloé. In the 1790s, ensign and head pilot of the Royal Armada, José Moraleda y Montero, charted the archipelago on orders from Peru's viceroy, Teodoro Croix, and the governor intendant of the Chiloé islands, Francisco Hurtado.[25]

It is worth noting briefly that the activities we are summarizing here were not carried out by individuals or isolated. Because being part of the viceroyalty's inner circle was the engineers' goal, they collaborated whenever possible with Sociedades Económicas y Patrióticas (Economic and Patriotic Societies); on the coast, with the help of consulates (*consulados*, the development policy advisory body for the last viceroys), viceroys encouraged the founding of nautical schools, institutions that had an educational function and served as true hydrographic depositories.[26]

No less important was the contribution these technicians made to the consolidation and coordination of geographic questionnaires, a method whereby they obtained massive amounts of information about the region. These questionnaires were administered during the eighteenth century with the collaboration of engineers and sailors. In 1743, the Consejo de Indias (Council of the Indies) urged the viceroy of New Spain, Pedro Cebrián, entrusted Mexico City chronicler Juan Francisco Sahagún de Arévalo (publisher of the *Gazeta de México* and the *Mercurio de México*) and the quicksilver accountant, Antonio de Villaseñor y Sánchez (author of a 1751 map of New Spain and of *Matemático cómputo de los astros* [Mathematical Computation of the Stars]) to draft the necessary questionnaires and to collect the reports to enable a better understanding of each jurisdiction in the viceroyalty. The questionnaires were about the location and distance between urban centers and their social and economic development. Antonio Ulloa, for example, as a commander of New Spain's fleet, in 1777 designed a plan for Viceroy Antonio María Bucareli to collect the scientific, historical, and statistic data missing from the 1743 geographical questionnaires.[27]

With questionnaire answers in hand, Villaseñor wrote *Theatro americano* (American Theater, 1748–1749), considered to be the first regional geography of México.[28] His contemporaries appreciated it greatly, and for a long time it was the compulsory starting point for new expeditions.

The constancy of expeditions of this kind throughout the colonial period, even after Spain started sending its own scientific expeditions, seems amply proven. Other examples from the Viceroyalty of Peru can be added to those we have mentioned from New Spain. The Consulado of Lima, between 1803 and 1805, sent questionnaires to its delegates and representatives in the provinces. The royal ordinance of August 25, 1802, was applied, which ordered the carrying out of geographic-economic surveys in all Spanish domains.[29] All the knowledge about nature and American culture the cosmographers and viceregal commissioners acquired was periodically reviewed and represented the core of works that synthesized the information, such as the *Diccionario geográfico-histórico de las Indias occidentales o América* (Geographical-Historical Dictionary of the West Indies or America), in the pages of which Antonio Alcedo y Herrera clearly expresses the importance of geography to the political administration of the empire: "It was ancient political dogma that seeing the kingdoms was an effective way of keeping them and that where monarchs could not see or be present, geographical demarcations and historical relationships took their place, since, just as the lenses of telescopes do, they make near and present the most distant objects."[30]

Besides the geostrategic dimension of viceregal interest in geography, the wish for regional promotion is also evident. Nowhere were the optimistic expectations for technology as keen as in the Sociedades Económicas de Amigos del País (Economic Societies of Friends of the Country).[31] These institutions coordinated the local administrative and technical elite and allowed the viceroyalty to capitalize on the scientific and reforming potential of recent graduates of the Convictorios Carolinos. Before such societies were founded, and even after they fell into crisis, there were remarkable relationships between the viceregal court and some renowned intellectuals.

A paradigmatic example might be miner, lawyer, writer, and scientist Joaquín Velázquez de León of New Spain. Between 1768 and 1770, he was adviser to *visitador* José de Gálvez during an assessment of the Californias. During this journey he made important astronomical observations, introduced remarkable improvements in mining, and made multiple suggestions for the economic development of that border territory. In 1771, he composed for Viceroy Carlos Francisco de la Croix a detailed report on his experiences, a report that would be decisive in the development of New Spain's mines.[32] The biography of Francisco Xavier Gamboa provides us with multiple examples that prove him to have been as influential as Velázquez.[33]

In Peru, the scientific activity of the last decade of the eighteenth century and that of the first two decades of the nineteenth was marked by the work of Hipólito Unánue, a man of the Enlightenment who did not hesitate to put his broad medical, botanical, geographic, and historiographic knowledge at the service of the projects of the Viceroyalty of Lima. He was a great supporter of the *Mercurio Peruano*, the primary organ of Lima's Sociedad de Amantes del País (Society of Lovers of the Country). He appears to have served as secretary to several viceroys, which we deduce from his preparation of *Guía política, eclesiástica y militar del virreinato* (Political, Ecclesiastical, and Military Guide to the Viceroyalty) at Viceroy Gil y Lemo's request, or his total or partial writing of the memoirs of Viceroys Gil y Lemos, Avilés, and José Fernando Abascal.[34]

Collaboration between Lima's cultural elite and the viceroy was decisive in founding the Anfiteatro Anatómico (Anatomical Amphitheater, 1792) and the San Fernando Colegio de Medicina (School of Medicine, 1808). The medical school was established only a few months after the publication of Unánue's most remarkable text, *Observaciones sobre el clima de Lima y sus influencias en los seres organizados, en especial el hombre* (Observations on the Climate of Lima and Its Influence on Organized Beings, Especially Man, 1805), and after Viceroy Abascal had appointed him director of the Protomedicato (Board of Medical Examiners) in November 1807.

During this period, New Granada enjoyed the services of an outstanding figure, José Celestino Mutis, the most important intellectual in the viceroyalty, who collaborated in all of its projects.[35] He arrived in 1760 as the viceroy's physician, and it did not take long for him to establish links with Bogotá's institutions. Thus, by 1762, he held the Mathematics Chair at the Colegio de Nuestra Señora del Rosario.[36] From then on, his desire to lead and administer would scarcely wane.[37] Always attuned to the viceroy of the moment, his social ascent was tied to reform projects or reports that he wrote at the regional authority's request.[38] There were few projects on which he did not work from the moment Viceroy and Archbishop Antonio Caballero y Góngora decided, in 1781, to create the Expedición Botánica and appointed Mutis director. Although the expedition would ultimately be formally incorporated into the expeditionary program promoted by the Jardín Botánico of Madrid, Mutis counted on an autonomy that was legitimized by the viceregal decision that botanical slides were not to leave Santa Fé de Bogotá and that the manuscripts were to remain with the viceregal secretary to later become part of the holdings of the Real Biblioteca (Royal Library) of the viceregal capital.[39]

In fact, the hypothesis that Mutis occasionally considered the expedition to be his personal property is well justified.[40] Not only did he try to promote numerous businesses using his position, but he also made his three nephews his only assistants and obfuscated reports of the work under his command. Among the numerous commissions he carried out for the viceregal authority were his endeavors to work gold deposits in Mariquita, his diverse projects to reform university education, and his participation in efforts to improve health care and in the fight against smallpox.[41]

In the Viceroyalty of Río de la Plata, the most outstanding character was army engineer Félix de Azara. In 1781, he arrived at Río de la Plata as a member of the boundary commission, which was charged with correcting the American boundaries of Spain and Portugal after the Treaty of San Ildefonso was signed (1777). For over a decade, he participated in the geographic demarcation of Brazil and in the definition of the boundaries between the Paraná River and Matto Grosso.[42] In the 1790s, he was in Buenos Aires and was placed in charge of a detailed exploration of all the southern borders. Later he visited the Spanish possessions south of Río de la Plata and Paraná. As a result of those commissions, Azara produced numerous maps: of the City of Corrientes, the Provinces of Misiones and Paraguay, and the course of the Paraguay River, for example. He also wrote several reports for the viceroy of Buenos Aires, including *Memorias sobre el estado rural del Río de la Plata* (Report on the Rural State of Río de la Plata), in which he analyzes the economic situation and the property patterns of agriculture in the province.[43]

The Expeditionary Tradition of the Home Country

The discovery of America had extraordinary scientific and cultural consequences. The firm belief that there remained no territories to be discovered made the world less mysterious and no longer a boundless space; it came to be seen as complete and therefore susceptible to order and measurement. This expectation immediately affected cartographic, botanical, and astronomical knowledge. After exploration and inventories were completed, it became necessary to systematize all data according to criteria that allowed encompassing the great diversity of phenomena and natural beings. It was an effort that required observations made according to a routine that improved over time and that became a specialized language.

At the same time, another factor became necessary for the worldwide spread of the scientific method: international research and, as a consequence, the strengthening of bonds between scientists and the seat of political power. Thus, Renaissance cosmopolitanism became scientific internationalism, which was institutionalized in centers that could coordinate a worldwide net of scholars and experts and that developed an innovative communication model using another language, new agents, different methods, and different legitimating procedures. Within such (scientific) communication systems, all participants historically tend to polarize around a center and a (paradigmatic) research program.

It is widely acknowledged that, during the seventeenth century, Spain lagged in the process that led to the Scientific Revolution. This circumstance pushed all scientific activities developed in the empire toward the periphery. Along with the internal processes of cultural renovation (among which those related to population and defense stand out) Spain's participation in international cooperative observation projects had a catalyzing effect on the reforms devised by the new Bourbon bureaucracy.[44] The integration of peninsular people and institutions into the international scientific network was particularly significant to those programs that related to scientific expeditions to America.[45]

The magnitude of these projects required diplomacy in addition to significant economic investment, all of which meant advancing the increasingly crucial relationships between the state and science. Furthermore, the geostrategic and economics aspects—philanthropy and business—were hopelessly entangled. As a consequence, some Spaniards and Hispanic Americans who espoused Enlightenment ideals had the opportunity to meet renowned European scientists and learn the accepted methods of observation and instrumentation.[46]

Our purpose is not to describe the activities of these expeditions, for this has been done by other scholars. Our interest here is to point out the legitimating character of these expeditions for the values and knowledge that would become decisive identifiers of outstanding representatives of the local elite, both in Spain and in Hispanic America. One of the cultural expressions of this fact was the reappraisal of the continent's economic possibilities after its natural resources were explored. Thus, along with the reasons why scientists were sent, we find observations and suggestions inviting the government into profitable business ventures and new sources for local prosperity, such as harvesting cinnamon in Venezuela, cinchona in Loja, cochineal in California, or wood in Guayaquil.

This consideration can be extended to all botanical expeditions sent from Madrid to Peru, New Granada, and New Spain between 1777 and 1787. Planned for the purpose of inventorying American flora, transplanting vegetable resources useful for the pharmacopoeia and gardening to Madrid's Jardín Botánico, and spreading the Linnaean classification model of nature, the intellectual concerns and political interests of the different power groups active in Hispanic America became intertwined insofar as organization, development, and results. We have already mentioned the circumstances that led to expeditions to Peru, New Granada, and New Spain;[47] another factor emerged with the creation of new institutes intended to promote the knowledge, methods, and values that met the home country's criteria, as happened with the establishment of the Cátedra de Botánica (Botany Department) entrusted to Cervantes, or with the introduction of the new chemical nomenclature in México,[48] two examples that historians recall to show the vigor of scientific Creolization in New Spain. In Lima, the presence of Creole scientists was rather small, and it was difficult for agents of the home country to institutionalize the new knowledge; in fact, they failed to impose new metallurgical knowledge in the Tribunal de Minería (Mining Tribunal) and to establish a botany department at the Universidad de San Marcos, in spite of the efforts made by religious and corresponding member of Madrid's Jardín Botánico, Francisco González Laguna, and by the military pharmaceutical attaché to the viceregal expedition to Peru, Juan José Tafalla.[49]

The expedition classifications that have been published and that follow disciplinary, administrative, and geostrategic criteria do not account for all the activities carried out on those expeditions.[50] In order to do that, we must first take into account the variety of responsibilities taken on by the participants and also the synchronicity of their work in America.

All the activities could be grouped according to three criteria, the functionality of which is determined by their power to distinguish between the most institutional, taxing, and home country–oriented behavior of these agents of the homeland, and service on some other commission by viceregal order, or, alternatively, cooperation, proselytism, and relinkage to the local elite.[51]

By looking at the expeditionary program, we can verify that the home country and institutional aspect of each were important but not the only thing. It is not enough to observe just one expeditionary function to understand the key elements of expeditionary actions and to evaluate their results. We must, according to our criterion, see the totality of the work they did in order to understand their numerous effects and the reasons for the sincere support or the severe criticism of local groups, or how those attitudes might have changed in the local elites through the years.

It is also interesting to note the synchronicity of much of the expeditionary movement and the simultaneity of the activities of different expeditions. This is an issue of great importance, since most activity—except in Mexico—took place in areas where the population was small and where, obviously, the presence of a dozen persons endowed with great political and scientific authority must have had a great impact. If we emphasize this point, the feedback and continuous nature of expeditionary policy is apparent, since the arrival of successive, overlapping waves of scientists and technicians made possible the transfer of science and technology between social groups. This becomes clearer if we remember that several explorers, among them, Cervantes, Longinos, Mutis, Bustamante, and Haenke, remained in America after finishing their missions, or that several Creoles, for instance, Mociño and Zea, traveled to Spain, where they held important positions.

Whatever the importance of such expeditions to scientific development in Europe, what we want to highlight here is their participation in the processes of regional integration and differentiation in the colonies. It is well known that they did not always have their operational quarters in the viceregal capitals but in more remote places, where the richest flora were found. Such places and their areas of influence were economically appraised. This happened in the central Andes, where plants had been collected by members of the botanical expedition to Peru since 1780, the year the Huánuco base of operations was established. Although far from Lima, Huánuco's location was privileged.[52] The area became much more important after Manuel Alcaraz—who had already visited the cinchona production zone in the Loja mountain range—discovered near Huánuco an abundance of that valuable tree. Several trained workers were brought from Loja to teach the natives how to gather the bark so that a profitable business could be set up in Lima. The arrival of Spanish botanists had an immediate effect on the development of the trade in cinchona as they helped differentiate the species and obtain the quinine extract.[53]

While staying in the Andes, Spanish botanists also took an interest in agriculture, particularly in Andean coca, since the miners of Cerro de Pasco were supplied from Huánuco. Their knowledge of the role of this plant in the religion, diet, and economy of the natives would change their European prejudices regarding this so-called vice plant: "The experience made me change," Ruiz writes, "showing me, with positive facts, the wonderful effects of those leaves, seemingly tasteless, odorless, and inert."[54]

European botanists therefore began all kinds of assignments in several parts of the viceroyalty. In early 1779, Viceroy Manuel Guirior entrusted French naturalist Dombey with an analysis of Chauchín's mineral waters, near the source of the Huaura River.[55] Dombey was also commissioned by the president of the Audiencia in 1773, during the naturalist's stay in Chile, to study the mercury mines in the north of that country. Pavón was very interested in carrying out the commissions entrusted to him to study the woods of the Chilean forests. And Tafalla would start a similar task in the Ecuadorian forest at the mouth of the Guayas River and in the Intag Mountains, where he would meet with Caldas.[56]

During the Malaspina expedition's stay in Lima (May–September 1790 and July–October 1973), both expeditions cooperated closely. Thus, Viceroy Gil y Lemos accepted Malaspina's suggestion that naturalist Juan

Tafalla and painter Francisco Pulgar participate as experienced guides in the scientific expeditions that Haenke, Née, and Pineda were to carry out in the Andean region.[57]

This kind of collaboration also existed among renowned members of the local elite: clergymen who cared for the dying González Laguna participated in natural history projects, and Father Francisco Romero helped with astronomical and geodetic tasks. González Laguna and Romero were suggested by Malaspina to work with Manuel Lavarden in Buenos Aires, Joaquín Toesca in Santiago de Chile, and others from the new patriotic society in Quito to create a meteorological correspondence network among the different Hispanic American cities and the Academia de Guardiamarinas (Coast Guard Academy) in Cádiz.[58]

It seems clear that these Spanish explorers were active participants in the cultural life of the region. Botanist Pavón, for instance, was praised by the local authorities for having directed the construction of a public boulevard in Huánuco.[59] They frequently had contact with the Franciscans from the Ocopa monastery; it is also possible that artists from the Gálvez and Brunete expedition collaborated in Martínez Compañón's iconographic album. Years later, Tafalla was the link between Lima and the Quechua philological work of clergyman Joseph Manuel Bermúdez of Huánuco; González Laguna encouraged this work, since Europe was interested in American languages.

We can say much the same about the activities of members of boundary-mapping expeditions or about Malaspina and his travels through Buenos Aires. Many of these explorers settled in Río de la Plata, for example, frigate captain Francisco Javier de Viana; a sailor, Juan Antonio Gutiérrez de la Concha, who explored the San Jorge Gulf and governed Córdoba de Tucumán; and the chief of Río de la Plata's naval station, but based in Montevideo, José de Bustamante y Guerra. Bohemian naturalist Tadeo Haenke settled in Upper Peru (Cochabamba), from where he sent articles to the Río de la Plata press and Cabello's *Telégrafo Mercantil*, to Vieytes and Cerviño's *Semanario de Agricultura* (Agriculture Weekly), and to Manuel Belgrano's *Correo de Comercio* (Business Mail).[60]

The members of the boundary commissions also participated in local life. Diego de Alvear, married to Josefa Barbastro (an important name among Buenos Aires merchants), handed in the instruments of his expedition to the consulate.[61] Pedro Cerviño, an engineer and member of the third boundary commission expedition, founded the Escuela de Náutica of Buenos Aires, where a select group of technicians was trained, for example, Bernardino

Rivadavia, Francisco de la Cruz, Lucio Mansilla, and Benito Goyena,[62] and where pilot and geographer Andrés de Oyarbide, member of the second boundary commission expedition, was also an adviser.[63]

The expeditions to New Spain were different from the rest that were sent to Hispanic America. In Mexico, a group of Creole intellectuals thought they were competent enough to be in charge and direct all projects charged with learning about the reality of New Spain. Sessé's instructions were to found a botanical garden and to create at the university in the viceregal capital a botany department wherein the Linnaean classification system would be taught. Both endeavors were part of a plan to reform the medicine curriculum and to ensure, as in Spain, that physicians' power on the Protomedicato would diminish. The controversies aroused by these initiatives have been recounted many times, and it could be said that they intensified debate and deepened the Creole elite's discontent after the expulsion of the Jesuits in 1767. Church elder Alzate, for instance, was accused of calling the Linnaean system nonsense and childish things, and urged botanists trained in Madrid to imitate ancient Mexicans.

When their stay in Mexico came to an end, the expeditionaries had accepted Alzate's opinion of the Linnaean system to a certain extent; Sessé, Mociño, and Montaña started a vast experimentation program to define the therapeutic properties of plants and to elaborate a Mexican materia medica. Ten years later, the alliances had been modified: the university, which had been a Creole bastion,[64] accused the expeditionaries of risking the lives of the patients at Hospital San Andrés and Hospital Natural and asked the viceroy to cancel the research. This subterfuge left the university socially isolated and without the support of some of its most renowned professors, for Profs. Joseph Gracida and Daniel O'Sullivan had decided to study with Cervantes and learn the new botanicals.[65] Thus, not only had the Spanish commissioners' crudely domineering ways been corrected but a more cooperative attitude had emerged.[66]

The aforementioned subterfuge might represent what also happened during the mineralogical expedition of the Señores Elhúyar, as Alzate called them, which gave rise to initially bitter debates brought about, according to Peset,[67] by the encounter of great European science and great New Spain technique.

We also find examples of the catalyzing capacity the expeditions had on Creoles in New Granada, the heart of the spread of modern science.[68] Independently from his commissions for the viceroy, Mutis collected and classified flora. It is no accident that his expedition is still described as botanical. In fact, he even imagined the possibility of correcting Linnaeus and, whether he was qualified or not for such an intellectual adventure, the truth is that all of his work was oriented toward a theoretical objective.

As a consequence, two crucial aspects of the expedition were neglected, one relating to methodology and the other to its objectives. The task of collecting plants, because of the sedentary character with which Mutis had endowed the expedition, was entrusted to envoys with almost no scientific education, hence the importance painters had in this expedition. Mutis's plan (unlike what happened in New Spain) also was unable to accomplish the utilitarian aims of the study of the flora. Both failures were strongly criticized by Creoles once Mutis lost some of his viceregal support. Caldas criticized Mutis's excessive iconicity: "Such grandiosity..., such literary luxury, contributes little, and, truly speaking, it delays the progress of the sciences." Expedition zoologist Tadeo Lozano complained that Mutis's theoretical insistence "tied [his, Lozano's] hands and prevented [him] from publishing as he went along."[69] Zea summed up all this criticism in *Proyecto de reorganización de la expedición botánica* (Botanical Expedition Reorganization Project, 1802) by stating that his objectives had been "purely botanical, with no relation whatsoever to agriculture, economics, or the arts."[70] These well-justified observations do not express any reality other than the one derived from the newly achieved maturity of New Granada's elite, which had not existed twenty years before and which had been able to take advantage of the contact with European science provided by the simultaneous presence of Mutis, Elhúyar, and Humboldt.[71]

The collaboration was mutual,[72] as shown by consulting the *Mercurio Peruano*, a journal that devoted a great many pages to learning about the country (33.8 percent) and to the spread of science (25.5 percent).[73] Ruiz, for instance, came to be considered the Peruvian Linnaeus, and González Laguna expressed pride in the new intellectual atmosphere in Lima: "Before this time we knew only the native [plants] of this country, very few from Spain. Today, we have [plants] from very remote parts of our continent."[74] Mociño acknowledged the benefit of his contacts with the New Spain expedition as he declared their studies on materia medica to have been carried out "not only as mere compilers but as exact observations."[75] The Lima press paid great attention to the labor of the mineralogists and members of the Nordenflicht expedition sent by the home country to study Andean mineral deposits and to introduce a new amalgamation method to replace the traditional patio process.[76] This simultaneity of scientific intervention in Peruvian territory made it possible for the members of the viceroyalty's local elite to foster the illusion that it was possible to restore Peru, thanks to the cultivation and development of modern science.

American Mosaic

Having presented in outline form the three scientific traditions that developed on the Hispanic American scene (those of the home country, the viceroyalty, and the Creole elite), we would like to stress that their interaction grew deeper with the arrival—almost in waves—of expeditions sent from Europe at the end of the eighteenth century. The expeditions, therefore, represent an extraordinary catalyst for cultural life in Hispanic America. Endowed with great authority, both political and scientific, and after proving their competence during the assignments local authorities entrusted to them, the expeditionaries became magnets for a variety of controversies and debates.

Where the maturity of Creole and viceregal local elites was more noticeable, the interactions were more fruitful. It is within this environment that cultural differentiation solidified so acutely that it was decisive in the process of cultural regionalization of the Spanish colonial system in America. This differentiation spread through the four scientific and political activity centers: Mexico City, Lima, Bogotá, and Buenos Aires, that is, the capitals of the four viceroyalties into which the American portion of the Spanish Empire was divided in the eighteenth century.

The aforementioned cultural regionalization must not be seen as a by-product of the colonial armies' reform projects, the implementation of intendancies, or the introduction and putting into practice of free trade. Such policies lasted only a short time and, on the whole, were fragmentary and unsuccessful. By cultural regionalization we mean the capacity of the local elite to find symbols that could replace reality and make it meaningful. Most significant at the end of the eighteenth century is the fact that such symbols were described, related, and idealized by incorporating the methods, language, and practices of modern science. This revealed the importance of science, as a substantial—and not merely instrumental—element in the Hispanic American elites' culture.

The regionalization process, therefore, expresses a dual maturity in Hispanic American culture. On the one hand, those who produced this maturity were able to invent a tradition and a portable symbology, with their integrating potential. On the other, the local elite was able to change that tradition and integrate into it the concepts and key institutions of European culture.

We hope to describe the key points of that process in a future study. We believe that this will be possible for the four cases we have mentioned here, keeping in mind that the vitality and novelty of the process are not legitimated by the elite's capacity for reproducing emancipating formulas comparable to the French republicanism of that historical moment; rather, the locally created symbols are the cultural elements that had a more corrosive, disintegrating, and revolutionary effect. Thus, while the elite of New Spain agreed to make Mexico City a universal metropolis and a splendid synthesis of the viceroyalty, with the Virgin of Guadalupe as its most detailed symbol of integration, in Peru, it would be the mystery of its sky. With an atmosphere and climate that were capable of integrating the national being and of explaining all of its amazing peculiarities, if—as Galileo explained—religion taught one how to go to heaven and science taught how it was made up, the Peruvians wanted to risk the belief that the route to heaven followed the trace the movement of its stars left behind. If in New Granada the idea of America as a Promethean environment enlivened the patriotic aspirations of its elites, in Río de la Plata, the dream of taming the inner Mesopotamia took on the function of mobilizing public opinion.

We are speaking, though, of elites, not so much for their education or wealth as for their capacity to articulate legitimating discourses that were, conversely, legitimated by the structures of administrative, political, and economic power in the colony: church, home country, and the viceroyalty. It is in this sense that we believe it necessary to consider the assignment of intellectuals to specific roles and to introduce a third factor—political and cultural dynamization—besides the Creole (Spanish American) and the peninsular (European Spanish). We believe this approach has the advantage of not being obsessed with racial features and, at the same time, of acknowledging the three power sources' capacity for political and cultural action.

Notes

1. The authors wish to thank the Dirección General de Investigación Científica y Técnica (Spain), grant no. PB91-0071. We also want to thank Juan Pimentel for his unselfish help and valuable advice.

2. Bauer, "The Church"; Tibesar, "The Peruvian Church."

3. Macera, "Iglesia y economía," p. 146; Brading, "Government," p. 402.
4. Brading, "Government," p. 398.
5. Mestre, *Despotismo*.
6. Pérez-Rioja, *Proyección*.
7. Vargas Alquicira, *La singularidad novohispana*.
8. Undoubtedly, much of these scientific enterprises' success was due to the incorporation of technological innovations into mining and agriculture. One of the most spectacular cases took place in the Paraguayan *reducciones*, where the Jesuits tried to acclimatize, develop, and monopolize the commerce in mate tea, thus reaping great economic benefits. They also introduced cotton and tobacco cultivation, as well as the use of iron instruments, an innovation that Roa Bastos ("Entre lo temporal," pp. 27–28), following Métraux, has described as "the axe revolution."
9. Sainz Ollero, *José Sánchez Labrador*.
10. Bitterli, *Los "salvajes,"* pp. 294–298.
11. Gerbi, *La disputa*, p. 212.
12. Roig, *El humanismo*, vol. 1, p. 88.
13. The content of their writings clearly shows the nature of the intellectual project these exiled men had in mind. Velasco's *Historia del Reino de Quito*, e.g., was conceived as a legitimating tool for the mestizo autonomist regional movement (Roig, *El humanismo*, vol. 1, p. 78). The work is divided into three parts: natural history, ancient history, and modern history. The first encompasses the mineral realm, the plant realm, the animal realm, and the rational realm; the second focuses on the primitive inhabitants of Quito until the Spanish conquest and then until 1767. The work shows a progression from the inorganic to the organic and from there to what is properly historical: human beings and their culture. To Velasco, "nature does not leap," and humankind, although placed at the summit of creation, does not separate from nature (ibid., pp. 97–98).
14. Arbesman, "Contribution," pp. 393ff.
15. Pérez Ayala, *Baltasar Jaime Martínez Compañón y Bujanda*, pp. 20–36.
16. Brading, "Government," p. 399.
17. The case of José Antonio Birt may be representative. After disembarking in Cumaná in 1754, he worked in Venezuela as a topographer to chart the seacoast, islands, harbors, and rivers from Cumaná to the Orinoco River, with the sole purpose of making commerce difficult for the Dutch, the French, and the Danish. In 1760, he was sent to Cartagena de Indias to collaborate in its fortification and was assigned to Panama by order of the viceroy of Nueva Granada to explore the Darien Gulf and improve its defenses. War was declared against England, and he was sent to Peru to help with the viceroyalty's defenses. First, he traveled to Valdivia in order to build a powder blockhouse and to improve the forts. Afterward, he went to Santiago and built the Dragoons' quarters; in Valparaíso, he focused on fortifications. After living temporarily in Lima, he made a new general inspection of the fortifications in Chile between 1768 and 1772, drawing new charts of Valdivia and the Juan Fernández Islands (Navarro Abrines, "Los ingenieros militares," pp. 57–58.).
18. Moncada, "Los ingenieros militares," p. 345.
19. Ibid., p. 320.

20. Bernabeu, "Las expediciones hidrográficas," pp. 353ff.
21. González-Ripoll, 1991.
22. Cook, *Flood Tide*.
23. Lozoya, *Plantas*.
24. *Exploraciones*.
25. Hanisch, *La isla de Chiloé*, p. 174.
26. Lima's Escuela de Náutica (Nautical School) was created in 1793, and seaman Balcato excelled there. The Buenos Aires school (1799) was founded at the request of the viceroy's technical counselors connected to the consulate, such as Belgrano and Azara. The school was headed by a geographer and cartographer, Pedro Cerviño; it closed in 1806 under pressure from the home country, as it was considered to be a "mere viceregal luxury" (Babini, 1986, pp. 46–47). In Cartagena de Indias, the Escuela de Náutica started in 1810. These institutions inherited the legacy of some viceregal functionaries who cultivated astronomical and physical geography, as had happened in Peru with the Cosmógrafos Mayores del Reyno (Elder Cosmographers of the Realm), an adjunct position to the Mathematics Department. Their work can be reconstructed, thanks partly to the uninterrupted publication of the *Conocimientos de los Tiempos* (Knowledge of the Times) series in Lima. Some of these cosmographers, like Koening, Peralta Barnuevo, or Godin, collaborated in the fortification of Lima; Cosme Bueno also handled the geographical descriptions that were sent by the magistrates from several places to create comprehensive descriptions of Peru (Macera, 1977).
27. Solano, *Antonio de Ulloa*, pp. cxliv–cl.
28. Solano, *Relaciones geográficas*, pp. 12–15.
29. The questionnaire's answers were reported as geographic accounts, or *razones circunstanciadas*, in which the consulate's power over Peruvian territory was evident. The consulate maintained exceptionally well organized files, a permanent source of information in its provincial delegates, and a group of counselors in the capital to whom technical, financial, and legal opinions were entrusted (Macera, "Informaciones geográficas," pp. 188–190).
30. Ibid., p. 182.
31. Schafer, *The Economic Societies*.
32. Moreno, *Joaquín Velázquez de León*, p. 63
33. Trabulse, *Francisco Xavier Gamboa*.
34. Woodham, *Hipólito Unánue*, pp. 228–230.
35. Restrepo, "El tránsito," p. 198.
36. Arboleda, "Mutis entre las matemáticas."
37. Peset, *Ciencia*, pp. 281ff.
38. Restrepo, "El tránsito," p. 194.
39. Peset, *Ciencia*, p. 286.
40. Amaya, *La Real Expedición Botánica*, p. 56
41. Quevedo y Zaldúa, 1988; Frías Núñez, *Enfermedad*.
42. Galera, *Félix de Azara*, p. 14.
43. Chiaramonte, *La crítica ilustrada*, p. 58.
44. Amaya, *Celestino Mutis*.
45. Spanish participation in the expedition to the Kingdom of Quito intended to determine the Earth's shape (Lafuente and Mazuecos, *Los caballeros*), or in the

one headed by Abbot Chappe (Moreno, *Joaquín Velázquez de León*) to California in order to observe Venus passing in front of the sun, had a similar origin: the definition of a problem by the Académie Royale des Sciences (Royal Academy of Sciences) and the Royal Society of London that required research that had to be carried out on the other side of the world. Although their objectives differed, the first expeditions to Hispanic America in the eighteenth century were related to problems of an international nature, as was the expedition to the Orinoco to establish the borders between Portugal's and Spain's domains (Lucena Giraldo, *Francisco de Requena y otros*). The botanical dimension of this expedition (entrusted to Petri Loefling, a pupil of Linnaeus)—as in the other cases mentioned—had been decided outside Spain. The same applies to the expedition to the Viceroyalty of Peru, part of a French endeavor led by Joseph Dombey and to which botanists Hipólito Ruiz and José Pavón were attached (Puerto Sarmiento, *La ilusión quebrada*). Previous expeditions were also international in nature; nevertheless, as Spanish scientific institutions matured, expeditions began to be financed as national enterprises to which scholars from other countries contributed.

46. Other contacts had already been established, such as the two trips Louis Fueillé made at the beginning of the century to South America, leaving his mark on the local scholars of Cartagena de Indias (Arias de Greiff, "Historia," p. 128) and Peru (Steele, *Flores*, p. 22). Feuillé's stays in Lima (in 1709) and, later, Frazier's (1712–1714) were different from others that would follow in terms of the capacity of colonial scientists to interact with European scholars. It is worthwhile to point out Pedro Peralta's frequent contacts with the Académie Royale des Sciences of Paris, an institution of which he would later become a corresponding member and to which he would promptly send his observations of celestial events.

47. Puerto Sarmiento, *La ilusión quebrada*, pp. 94ff.

48. Aceves Pastrana, "La difusión."

49. Notwithstanding what's been said, the Linnaean method spread throughout Peru thanks to the nearly missionary efforts of Ruiz and Pavón among Lima's scientific elite. Thus, during the early months of 1780, at the Universidad de San Marcos, they had the opportunity to perform before some members of the university an *objective lesson and demonstration* of their botanical and Linnaean skills. Unánue's introduction to the scientific description of Peru's plants, published in the *Mercurio Peruano* under the pseudonym Aristio, is also evidence of the enthusiastic reception of the Linnaean system.

50. Pino and Guirao, "Las expediciones ilustradas"; Lucena Salmoral, "Las expediciones científicas."

51. By working through institutions, agents showed an ability to impose the way scientific work should be organized, according to criteria from the home country. They made an effort to spread and reconcile European scientific practices. Thus they achieved—frequently through authoritarian methods—the introduction of the Linnaean system. By working through missionaries or cooperating with one another, they gained some scholars' participation. Thus, they fostered the construction, strengthening, and expansion of information-exchange networks, which tended, by their very nature, to be useful tools for the institutionalization of knowledge from the home country. In addition, they built around a center that was no longer in the home country but in the colony, which was its most characteristic and solid trait. In

performing commissions, the agents showed their ability to become part of viceregal projects and to be used as qualified technicians and sources of information by the local powers. Activities of this type were described in reports, judgments, accounts, that is, writings that increased the administrative literature generated by the colonial bureaucracy.

52. Huánuco had some six thousand inhabitants and was located five hundred kilometers northeast of Lima, on a forested slope of the eastern Andes, near the jungle. Nearby, the Huallaga River reaches the Marañón and the Amazon. Dombey qualified its position as the "nec plus ultra de la Conquista española en el interior" [best of the Spanish domnination in the countryside].

53. If, in 1779—the year before Ruiz and Pavón arrived in Huánuco—the annual raw bark harvested amounted to around two thousand or three thousand arrobas (1 arroba equals twenty-five pounds, so this was twenty-three thousand to thirty-five thousand kilos), in 1788, some forty thousand arrobas were estimated to have been sent to Lima (Steele, *Flores*, pp. 96–97). From 1779 to 1788, Ruiz and Pavón researched the chemistry of producing the quinine extract in order to improve the therapeutic virtues of the bark; this fact accounts for the increase in exports.

54. Ibid., p. 101.
55. Ibid., p. 89.
56. Estrella, "La expedición botánica," p. 56.
57. Higueras, *Catálogo crítico*, p. 122.
58. Ibid., p. 136.
59. Steele, *Flores*, p. 225.
60. Ratto, *Alejandro Malaspina*, pp. xxviii–xxix
61. Tjarks, *El Consulado*.
62. Ratto, *Alejandro Malaspina*, p. xii
63. Lucena Giraldo, "Ciencia," p. 172.
64. Tanck de Estrada, "Aspectos políticos."
65. Lozoya, *Plantas*, p. 128.
66. Saldaña, "The Failed Search," p. 47.
67. Peset, *Ciencia*, p. 176.
68. Silva, *Prensa*, p. 183.
69. Restrepo, "El tránsito," p. 214.
70. Ibid.
71. Puig-Samper, "La ciencia metropolitana."
72. On June 29, 1790, Malaspina sent the former and current trade consuls of Lima a list of the latitudes and longitudes the ships under his command had determined during their trip to Cape Horn (Higueras, *Catálogo crítico*, p. 129). During his second stay in Lima, Malaspina was required by the viceroy to send all the botanical works he no longer needed to botanists Juan Tafalla and Francisco Pulgar (ibid., p. 199).
73. Clement, *Índices*, p. 17.
74. González Laguna, "Memoria," p. 166.
75. Lozoya, *Plantas*, p. 177.
76. European mine owners' efforts to improve the Peruvian mining situation

and their various experiments at diverse mining sites to prove the superiority of their amalgamation method were published by *Mercurio Peruano* (vol. 1, pp. 218–220, 288–289; vol. 2, pp. 30–32, 53–55, 149, 266–275; 3, pp. 217–223, 225–229, 233–239, 241–253; vol. 5, pp. 35, 227–229; vol. 7, pp. 46, 49, 66–81). The support this modernization effort received was justified on humanitarian grounds, since the use of the patio process saved the *repasiri* Indians' feet from rough labor [*repasiri* are the workers who trample the amalgam in the patio method].

Bibliography

Aceves Pastrana, P. "La difusión de la ciencia en la Nueva España en el siglo XVIII: La polémica en torno a la nomenclatura de Linneo y Lavoisier." *Quipu* 4, no. 3 (1987): 357–385.
Amaya, J. A. *Celestino Mutis y la Expedición Botánica*. Madrid: Ediciones Debate/Itaca, 1986.
———. *La Real Expedición Botánica del Nuevo Reino de Granada*. Bogotá, 1982.
Arbesman, R. "Contribution of the Franciscan College of Ocopa in Perú to the Geographical Exploration of South America." *The Americas* 1 (1947): 393–417.
Arboleda, L. C. "Mutis entre las matemáticas y la historia natural." In *Historia social de las ciencias: Sabios, médicos y boticarios*, D. Obregón (ed.). Bogotá: Universidad Nacional de Colombia, 1987.
Arias de Greiff, J. "Historia de la astronomía en Colombia." *Ciencia, tecnología y desarrollo* 11 (1987): 119–162.
Azara, Félix de. *Descripción general del Paraguay*, A. Galera (ed.). Madrid, 1990.
Babini, J. *Historia de la ciencia en Argentina*. Buenos Aires, 1986.
Bauer, A. J. "The Church in the Economy of Spanish America: *Censos* and *Depósitos* in the Eighteenth and Nineteenth Centuries." *Hispanic American Historical Review* 63 (1983): 707–733.
Bernabeu, S. "Las expediciones hidrográficas." In *Carlos III y la ciencia de la Ilustración*, M. Sellés, J. L. Peset, and A. Lafuente (eds.). Madrid, 1988.
Bitterli, U. *Los "salvajes" y los "civilizados": El encuentro de Europa y Ultramar*. Mexico City, 1982.
Brading, D. A. "Government and Elite in Late Colonial Mexico." *Hispanic American Historical Review* 53 (1973): 389–414.
Buecher, R. M. "Technical Aid to Upper Peru: The Northernflicht Expedition." *Journal of Latin American Studies* 5 (1973): 37–77.
Chiaramonte, J. C. *La crítica ilustrada de la realidad: Economía y sociedad en el pensamiento argentino e iberoamericano del siglo XVIII*. Buenos Aires, 1982.
Clement, J. P. *Índices del* Mercurio Peruano, *1790–1795*. Lima, 1979.
Cook, W. *Flood Tide of Empire: Spain and the Pacific Northwest, 1543–1819*. New Haven, 1973.
Estrella, E. "La expedición botánica en el virreinato del Perú (1777–1815)." In *La expedición botánica al virreinato del Perú (1777–1788)*, vol. 1, A. González Bueno (ed.). Barcelona, 1988.

Exploraciones y actividades marítimas españolas en el litoral patagónico argentino durante los siglos XVII y XVIII, 4 vols. N.p.
Frías Núñez, M. *Enfermedad y sociedad en la crisis colonial del Antiguo Régimen*. Madrid, 1992.
Gerbi, A. *La disputa del Nuevo Mundo*. Mexico City: FCE, 1960.
Góngora, M. "Estudios sobre el galicanismo y la Ilustración católica en la América española." *Revista Chilena de Historia y Geografía*, no. 125 (1980): 96–151.
González Claverán, V. *La expedición científica de Malaspina en Nueva España*. Mexico City, 1988.
González Laguna, F. "Memoria de las plantas extrañas que se cultivan en Lima introducidas en los últimos 30 años hasta 1794." *Mercurio Peruano* 11 (July 10, 1794): 163–170.
González-Ripoll, M. D. "Las expediciones hidrográficas en el Caribe: El Atlas Americano." In *La ciencia española en Ultramar*, A. Díez Torre et al. (eds.). Madrid, 1991.
Hanisch, W. *La isla de Chiloé, capitana de rutas australes*. Santiago de Chile, 1982.
Higueras Rodríguez, M. D. *Catálogo crítico de los documentos de la expedición Malaspina (1789–1794) del Museo Naval*. Vol. 1. Madrid, 1985.
Lafuente, A. "Institucionalización metropolitana de la ciencia española en el siglo XVIII." In *Ciencia colonial en América*, A. Lafuente and J. Sala Catalá (eds.). Madrid, 1992.
———, and A. Mazuecos. *Los caballeros del punto fijo: Ciencia, política y aventura en la expedición geodésica hispanofrancesa al virreinato del Perú en el siglo XVIII*. Barcelona, 1987.
———, and J. Sala. "Ciencia colonial y roles profesionales en la América española del siglo XVIII." *Quipu* 6, no. 3 (1989): 387–403.
López, L.; J. Luque; and R. Alcalá. *Arbitrios técnicos de la minería colonial*. Lima, 1986.
Lozoya, X. *Plantas y luces en México: La Real Expedición Científica a Nueva España (1787–1803)*. Barcelona, 1984.
Lucena Giraldo, M. "Ciencia para la frontera: Las expediciones de límites españolas." *Cuadernos Hispanoamericanos* (special issue, Carlos III and America) (1988): 157–173.
———(ed.). *Francisco de Requena y otros: Ilustrados y bárbaros. Diario de la exploración de límites a las Amazonas*. Madrid, 1991.
Lucena Salmoral, M. "Las expediciones científicas en la época de Carlos III." In *La ciencia española en Ultramar*, A. Díez Torre et al. (eds.). Madrid, 1991.
Macera, P. "Iglesia y economía en el Perú durante el siglo XVIII." *Letras* (Lima) 1, no. 70–71 (1963): 118–159.
———. "Informaciones geográficas del Perú colonial." *Trabajos de Historia* (Lima) 1 (1977): 181–233.
Mestre, A. *Despotismo e Ilustración en España*. Barcelona, 1976.
Moncada, O. "Los ingenieros militares en América." In *De Palas a Minerva: La formación científica y la estructura institucional de los ingenieros militares en el siglo XVIII*, by H. Capel, J. E. Sánchez, and O. Moncada. Barcelona, 1988.

Moreno, R. *Joaquín Velázquez de León y sus trabajos científicos sobre el valle de México (1773–1775)*. Mexico City, 1977.
Navarro Abrines, M. C. "Los ingenieros militares del virrey Amat: Un apunte biográfico." In *Ciencia, vida y espacio en Iberoamérica*, vol. 2, J. L. Peset (ed.). Madrid, 1989.
Pérez Ayala, J. M. *Baltasar Jaime Martínez Compañón y Bujanda (1737–1797)*. Bogotá, 1955.
Pérez-Rioja, J. A. *Proyección y actualidad de Feijoo*. Madrid, 1965.
Peset, J. L. *Ciencia y libertad: El papel del científico ante la independencia americana*. Madrid, 1987.
Pino, F. del, and Á. Guirao. "Las expediciones ilustradas y el estado español." *Revista de Indias* 180 (1987): 383–394.
Puerto Sarmiento, J. *La ilusión quebrada: Botánica, sanidad y política científica en la España ilustrada*. Barcelona, 1988.
Puig-Samper, M. A. "La ciencia metropolitana y la conciencia nacional en las colonias." In *Ciencia técnica y estado en la España Ilustrada*, J. Fernández Pérez and I. González Tascón (eds.). Madrid, 1990.
Ratto, H. R. (ed.). *Alejandro Malaspina: Viaje al Río de La Plata en el siglo XVIII*. Buenos Aires, 1938.
Restrepo, O. "El tránsito de la historia natural a la biología en Colombia (1784–1936)." *Ciencia, Tecnología y Desarrollo* 10 (1986): 181–275.
Roa Bastos, A. "Entre lo temporal y lo eterno." In *Tentación de la utopía: La república de los Jesuitas en el Perú*, R. Bareiro Saguier and J.-P. Duviols (eds.). Barcelona, 1991.
Roig, A. A. *El humanismo ecuatoriano de la segunda mitad del siglo XVIII*. 2 vols. Quito, 1984.
Sainz Ollero, H., et al. *José Sánchez Labrador y los naturalistas de los Jesuitas del Río de La Plata*. Madrid, 1989.
Saldaña, J. J. "The Failed Search for 'Useful Knowledge': Enlightened Scientific and Technological Policies in New Spain." In *The Cross Cultural Transmission of Natural Knowledge and Its Social Implications: Latin America*, J. J. Saldaña (ed.). Vol. 5. Acts of the XVII International Congress of History of Science, Berkeley, California, 1985. Cuadernos de Quipu 2. Mexico City, 1987.
Schafer, R. J. *The Economic Societies in the Spanish World, 1763–1821*. Syracuse, 1958.
Silva, R. *Prensa y revolución a finales del siglo XVIII*. Bogotá, 1988.
Solano, F. de. *Antonio de Ulloa y la Nueva España*. Mexico City, 1979.
——— (ed.). *Relaciones geográficas del arzobispado de México, 1743*. 2 vols. Madrid, 1988.
Steele, A. R. *Flores para el rey: La expedición de Ruiz y Pavón y la flora del Perú (1777–1788)*. Barcelona, 1982.
Tanck de Estrada, D. "Aspectos políticos de la intervención de Carlos III en la Universidad de México." *Historia Mexicana* 2 (1988): 181–197.
Tibesar, A. "The Peruvian Church at the Time of Independence in the Light of Vatican II." *The Americas* 26 (1970): 349–375.

Tjarks, G. O. *El Consulado de Buenos Aires y sus proyecciones en la historia del Río de la Plata.* 2 vols. Buenos Aires, 1962.
Trabulse, E. *Francisco Xavier Gamboa: Un político criollo en la Ilustración mexicana.* Mexico City, 1985.
Vargas Alquicira, S. *La singularidad novohispana en los Jesuitas del siglo XVIII.* Mexico City, 1989.
Woodham, J. E. *Hipólito Unánue and the Enlightenment in Peru.* Ann Arbor, 1964.

CHAPTER 5

Science and Freedom: Science and Technology as a Policy of the New American States

JUAN JOSÉ SALDAÑA

Science, arts, culture, and all that makes the glory of men and excites their imagination on the European continent, will fly to America.
SIMÓN BOLÍVAR

Between 1810 and 1824, the movement for independence of the Latin American nations emerged and was consolidated. In most cases, independence was attained after a cruel and prolonged war against Spain.

Along with the armed revolution, another revolution took place in the heart of Latin American society, the result of transformative forces active since the end of the colonial period and others set in motion by the independence movement itself. This revolution was intellectual in nature and led to the conception of the full sovereignty of nations in the face of King Ferdinand VII's rights. It also allowed political objectives to take shape. Fulfilling those objectives would be the task of the new states. The constitutional history of Latin America is the expression of the new mentality that emerged in 1810 and in which science played a role that was little known until now.

As pointed out in the previous chapters dealing with the Scientific Revolution during the colonial period, the interest in science was the result of various factors, both external and local. The local factors (the vital element in these changes) stemmed from the social reform projects conceived by the Creole sectors, which were discontented with the isolation and exclusion Spain imposed on them. Among such projects were the mining reforms in Mexico, initially championed by Francisco Xavier Gamboa (in 1761) and later by Joaquín Velázquez de León and Juan Lucas Lassaga (1774) on behalf of the mining union; educational reforms in Lima, sponsored by José Baquíjano and José Toribio Mendoza (1786);

and medical and surgical reforms, by Hipólito Unánue (1797). Social advancement of Creoles and the development of a patriotic science were headed by naturalists Francisco Zea and Francisco José de Caldas in New Granada at the end of the century, and medical reforms were carried out in Quito by physician Eugenio Espejo, among others.

Acting through economic and "friends of the country" societies, or as individuals connected to social sectors that were interested in reform, scientists promoted patriotically oriented (that is, useful) science. Within this context, the dissemination of science in the newspapers created by the scientists themselves in Mexico and Lima in the beginning, and later in Havana, Santa Fe, Guatemala, Quito, and other capitals, was of particular importance. News about new theories and technical applications as well as results of research on Latin America's natural resources, geography, industry, economy, and population were printed in these publications. This helped socialize the reform project that underlay Enlightenment modernity and helped it spread throughout the continent. The birth of scientific nationalism could be discerned by this stage.

The crucial moment for adapting modern science to America came in the 1780s, when boosters of science found a social role for it. The alliances they established with several social sectors (mine owners, tradesmen, etc.) allowed for a quite successful institutionalization process. Among the new institutions that fostered practical physics, chemistry, astronomy, botany, mineralogy, medicine, and surgery were the Seminario de Minería (Mining College) (1792) and the Jardín y Cátedra de Botánica (Botanical Garden and Department) (1788) in Mexico, the Bogotá observatory (1803), and the Colegio de Cirugía y Medicina of San Fernando (College of Surgery and Medicine) in Lima (1815). Others, such as the project to create a public university in Santa Fe, were not viable in the colonial environment, with its centralized economic, cultural, and bureaucratic framework.

In a few years, the institutionalization of modern science and technology began to produce results. The first generations of scientists and technicians, who were fully aware of the value of science to them and their society, were trained. Research that had practical applications for Latin America's natural resources, industry, geography, economy, and population was carried out. Such studies had an immediate impact on sectors such as mining, agriculture, and textiles but, above all, contributed to demonstrating the strength of the Spanish colonies as independent nations.

As a consequence, at the beginning of the nineteenth century there was a movement for science and the "useful arts" in practically every region of Spanish America. A significant number of scientists formed a community in each country, and institutions especially dedicated to the fostering and teaching of science were established. Valuable results were achieved in fields such as chemistry, natural history, mineralogy, and astronomy. Among increasingly wider sectors of society, consciousness grew of the progress and social well-being that could be expected from science. Men of the time became enthused and envisioned a great future for their countries. New Granadan patriot and scientist Francisco José de Caldas wrote thus from his native Popayán in 1801: "I am certain that when the nineteenth century ends, we won't have to envy the home country for its Enlightenment."[1]

Independence movements began less than a decade later. Scientists, probably because of their contact with modern thinking and their newly acquired knowledge about Latin American reality, were, from the start, sensitive to and supportive of the ideal of freedom that motivated the insurgents. Many of them participated actively in the wars of independence, lending their knowledge in the wars of emancipation. Such was the case of Caldas, who took part in the conspiracy that started the rebellion of July 20, 1810, and who, as a scientist, lent great services to the war of independence as a military engineer and as editor of the first newspaper of the republic, the *Diario Político* (Political Daily). During the Spanish reconquest of 1816, he was executed with other patriots who, like him, had been connected with the Expedición Botánica (Jorge Tadeo Lozano and Lino de Pombo). Another example is a group of graduates of the Seminario de Minería in Mexico, made up of Casimiro Ramón Chovel, Isidro Vicente Valencia, José Mariano Jiménez, Rafael Dávalos, and Ramón Fabié, the last a Filipino, all of whom joined Miguel Hidalgo's troops in Guanajuato in 1810 and who, too, were martyred.

The birth of the new nations brought hope for adequately supported science, which had to overcome the lack of encouragement of the colonial regime. Also, Latin American scientists felt the time had come for them to realize their cognitive and social advancement ambitions, which had been postponed for so long. No less important was their intention, shared with other social sectors, to achieve the "public good" through science and the useful arts. In this way, the newly independent states, as part of the liberal reforms that guided their leaders, manifested their interest in the development of education, science, and technology and decided to rely on these to reach proposed social and political goals. Science and technology ceased to be essentially a private affair, as they had been during the colonial regime, and became an issue of public interest.

It is also interesting to note how scientists played a leading role in designing the new nations once the institutional phase began. Many of them were congressmen in the constituent assemblies and left their special stamp on legislative work, as it was generally recognized that science and education were significant in the development of the new nations. So it happened in Cundinamarca in 1812, with Jorge Tadeo Lozano and Camilo Torres; in Argentina in 1810 and 1826, with Bernardino Rivadavia; in Peru in 1822, with Hipólito Unánue, José Pezet, and Miguel Tafur, among others; and in Mexico in 1824, with Valentín Gómez Farías.

Some of these men rose to the highest offices in their countries and thereby contributed to the fulfillment of the republican project for which they worked and in which they created a place for science. In Colombia, the eminent zoologist and mathematician Jorge Tadeo Lozano collaborated with Mutis in the Expedición Botánica and was the first governor of the State of Cundinamarca. During his tenure, the first constitution was promulgated, which declared full independence and which included fundamental provisions for science and the development of education. In Peru, from the moment General San Martín entered Lima, he invited the distinguished naturalist and physician Hipólito Unánue to join the first independent government as minister. In 1826, Unánue was named minister by Simón Bolívar and later became chairman of the Consejo de Ministros (Council of Ministers). Under his direction, the Dirección General de Estudios (General Research Office) and the Museo de Historia Natural (Natural History Museum) were created. In Mexico, in 1833, distinguished physician Valentín Gómez Farías headed the liberal government that introduced important reforms in education, closing down the old Universidad de México and replacing it with modern scientific institutions dedicated to the study of natural and physical sciences, geography, and statistics, among other subjects.

An issue that has not been adequately addressed in spite of its relevance is science's inherent relationship to the concept of Latin American nation-states. Public policy concerning science and technology was of crucial importance and helped determine the new countries' legitimacy. Science and technology became, in fact and as a matter of law, the preferred vehicle for Latin American nations to instill the notion of social equality through education, to create and develop an economy that could overcome the colonial bureaucratic and centralist vices that were so deeply rooted and harmful, and to provide the nation-state with the necessary means to justify its power rationally.

Latin American constitutions contain similar provisions regarding matters such as education, science, and technology. The sovereignty of states as established in the constitutions lies in the hands of the people and is articulated through nationally representative bodies like the Congress, the Assembly, and so on, and the states are responsible for the development of education, science, and scientific institutions; technology was seen as the means to accomplish the common good. All of this was put down in writing in, among others, the Constitutions of Caracas of 1810, of the New Granadan provinces (e.g., Cundinamarca, Tunja, Antioquia, and Cartagena) in 1810, of Mexico in 1814 (Apatzingán) and 1824, of Bolivia in 1826, and of Argentina also in 1826.

Among the first acts of the Latin American freedom movement was the establishment of a constitutional state in tune with the liberal, philosophical, and political thought of the time (Machiavelli, Bodin, Bacon, Hobbes, and Montesquieu, among others). As a consequence, the constitution of each country includes the nation's political self-definition and provides continuity in the historical national project, in spite of variations created over time.

Before analyzing some of the constitutions, I must point out that the economic sector's demand for science was, for all intents and purposes, absent during the initial stages of national life and subsequently. This stemmed from the features of economic organization—centralization and inefficiency—inherited from colonial times, which delayed modernization. These facts distinguish the difficult circumstance in which the Latin American nations came to independent life. If, in addition, we take into account the unstable conditions caused by the wars of independence and the reconquest and those caused by internal struggles for power during the early republican stage, we find that the resulting picture determined how Latin American science was structured and organized. In fact, the state was the factor that furthered and organized scientific and technological activity during the founding of the nations of Latin America. It placed itself in charge of national scientific activity for ideological and political reasons consistent with the modern philosophical-juridical doctrine on which it was based and also because no other social sectors were interested in science.

In Cundinamarca and in what later became the territory of Gran Colombia (Ecuador, Venezuela, and Colombia), this line of thought was expressed in the Constitution of Bogotá promulgated in 1811 by the Colegio Constituyente y Electoral (Constituent and Electoral College). This document declared complete independence from Spain and stipulated that every institution dedicated to education, industrial development, commerce, and the like was to remain under the executive branch so as to accomplish the "common good." Article 11 addresses public education, pointing out the state's obligation to create a patriotic society, which would foster "the fields of science, agriculture, industry, trade, factories, arts, commerce, etc." It also orders the strengthening of the Expedición Botánica and the teaching of natural science as well as the creation of a public university and secular schools.

In Mexico, the first constitution was proclaimed while José María Morelos was fighting against Spanish troops. The Decreto Constitucional para la Libertad de la América Mexicana (Constitutional Decree for the Freedom of Mexican America), or the 1814 Apatzingán Constitution, establishes the Mexican state's dominion by asserting that "sovereignty lies originally in the people and its practice by national representation" (Art. 5). Like other Latin American constitutions, the Apatzingán Constitution establishes that "education, as it is necessary to all citizens, must be favored by society with all its power" (Art. 39). As for science and technology, it establishes the following: "No kind of culture, industry, or commerce can be forbidden to the citizens" (Art. 38). It points out the state's obligation to "mint currency, specifying its material, value, weight, kind, and denomination; and to adopt the system of weights and measures the state considers as just" and also "to favor all branches of industry providing the means for its progress and to look after the people's enlightenment with particular care" (Art. 117). It also claims as the state's duty "to approve all regulations aimed at the citizens' health" and "to attend to and take care of all workshops and manufacturing of guns, cannons, powder factories, and the construction of all kinds of military hardware" (Art. 161). This Constitution, drafted by a Congress that had to change its location several times because of the war, was in force for several months in the territories held by the revolutionaries. Nonetheless, it is considered to be the original source of Mexican law and sovereignty.

The 1824 Mexican Constitution was drafted once independence and the republic were consolidated. Enlightenment and liberal thought of the time imposed their characteristic mark, assigning the state responsibility for the education of its citizens and fostering science and industry. The laws, decrees, and regulations that were issued during the early federal republic (1824–1836) reinforced the state's responsibility in these areas.

The authors of this Constitution acknowledged prescriptions stemming from a modern doctrine that contained "the bases of human associations

[and] the lost rights of humankind." This is why they assert the following in the Constitution's preamble:

> And as Mexicans opened their eyes to the light these principles cast, they declared that neither force nor interests nor superstition would rule their government. They have said, as a philosopher-writer did, that, after finding out the secrets of nature with Newton; after defining and setting forth the principles of society with Montesquieu and Rousseau; after extending the surface of the known world with Columbus; after snatching the thunderbolt from clouds to give it a direction with Franklin; after giving man's productivity indestructible life and unlimited range with other creative minds; finally, after linking all men through a thousand bonds of commerce and social relations, they can no longer tolerate anything but governments in agreement with this order, which has been created by so many and such precious achievements.

The explicit and revealing reference in this text to scientists of the Scientific Revolution, such as Newton, and to North American scientists, such as Franklin, as well as to theoreticians of the modern state, implies an intention to imitate them. The creators of the nation-state intended to make in Mexico a political revolution on the same scale as mentioned in the foregoing quotation and with a similar rational orientation. The arrangement of the country's political and social affairs was intended to be analogous to the newly established order regarding natural and social science. This was the civilizing task the state took on on the orders of its national representatives and is why it was said later, in the Constitution's preamble, that "this new order demands from us great sacrifices and a religious respect for moral values." In fact, the framers claimed that, with regard to the philosophical-political conception that guided the constitutional assembly's work, the political task was ethically required to stimulate "education of the young" and to follow the constitutional prescription of "fomenting the Enlightenment" (Art. 50, sec. I). They also warned "children of our Scholastic education" that the preamble meant that no evasion or subterfuge would be applied to elude these constitutional precepts; otherwise, "let us now renounce the right to be free."

The Constitution enumerates, in the articles addressing the exclusive powers of Congress and the states' responsibilities, the state's scientific and technological jurisdiction, that is, spreading the Enlightenment ideal; creating institutions dedicated to teaching science, the arts, and languages; creating military schools; protecting copyrights and patents; stimulating engineering works in the public interest; setting up a uniform system of weights and measures; granting awards to great men and for patriotic services; reporting by states on their agriculture, industry, and commerce and the means to develop them.

The states' efforts to bring to life or to sustain and encourage scientific and technical works were numerous. Heading the state of Gran Colombia were Simón Bolívar and Francisco de Paula Santander, who encouraged the creation in Bogotá of a museum, a botany department, and a school of mining. Modern chemistry and mineralogy were to be taught in the last, so Alexander von Humboldt and botanist Francisco Antonio Zea gathered a number of European and American scientists. In fact, the fostering of science and technology became policy, as shown by the Ley de Instrucción Pública de la Gran Colombia (Public Education Law of Gran Colombia) of March 1826: "Each university must have a public library, a natural history department, a chemistry laboratory, a botanical garden with the necessary assistants," and the corresponding professorships (Art. 35).

Other countries followed similar procedures; Argentina had had only minimally developed science up to that time, but with the revolution of May 1810, the Junta de Gobierno (Governing Junta) indicated the need to create an institution in which to train the young and in which the wise men and the patriots of the country would actively participate. The following are examples of the projects that were carried out: in 1810, Manuel Belgrano, a member of the junta, created a mathematics school that was financed by the Consulado (Consulate) (the organization for tradesmen in Buenos Aires). Shortly thereafter, the Directorio (Board) created the Academia Nacional de Matemáticas (National Mathematics Academy), financed by the government. In Mendoza, José de San Martín ordered that the Colegio de la Trinidad be created, where physics, mathematics, geography, and drawing, among other subjects, would be taught. Bernardino Rivadavia, as a member of the Junta de Gobierno and later as president of the republic, pushed for the institutionalization of science. His actions led to the creation of various institutions in which modern science and medicine were fostered and resulted in the creation of the Universidad de Buenos Aires in 1821.

In Peru in 1820, San Martín proclaimed independence and assumed command with the rank of protector. He immediately ordered that the Colegio de Cirugía y Medicina of San Fernando, which had been at the heart of scientific renewal, should henceforth be named the Colegio de la Independencia (Independence College).

In Bolivia between 1825 and 1827, Antonio José de Sucre reformed higher education as he organized the new state. He ordered the creation of colleges of arts and sciences in the major cities, and, to work "for the progress and for the teaching and dissemination of science, arts, and literature," the Instituto Nacional de Artes y Ciencias (National Institute of Arts and Sciences) in 1827. Around the same time, the teaching of modern medicine was instituted in Chiquisaca, Cochabamba, and La Paz. Also in 1827, the teaching of Tracy's, D'Holbach's, and Bentham's doctrines was made compulsory so as to fight Scholasticism and religious fanaticism with the rational, material, and utilitarian theories of the eighteenth and nineteenth centuries. No less important was the educational reform carried on by Santander in Colombia, following Bentham's ideas.

The reform of the universities that were inherited from colonial times was also a part of the new national governments' policies. In Caracas in 1827, new university bylaws were put in place with the participation of physician José María Vargas. Among other things, they stipulated the abolition of the racial, social, and religious discrimination that regulated acceptance to university—such as the "pure blood" requirement. Academic departments, particularly medicine and mathematics, were also reformed, and the universities were given their own funding.

Bolívar imposed similar reforms in Quito, Trujillo, and Arequipa. In Bogotá in 1828, the Colegio del Rosario and the Colegio de San Bartolomé were again authorized to provide higher education, and professors were asked to write university texts for their subject areas. A school of navigation was created in Guayaquil.

In Mexico, the government's scientific and technological policies were driven by the same spirit that permeates the aforementioned constitutional texts. The government's main task was to educate the citizens and to train the technicians the republic needed. It also had to ensure, through science and technology, the public good. Several decrees relating to education were issued between 1824 and 1826 that ordered changes and improvements in elementary public education; updated medical education; support for the teaching of science (in the Jardín Botánico and the Seminario de Minería), technology (in the Academia de San Carlos), and military and navigation instruction; and the creation of new professions, such as geographical engineer. The creation in various cities of patriotic societies that would help develop industry was ordered in 1822. In 1828, the Museo Nacional (National Museum) was authorized.

Almost every governmental action was marked by the use of science and technology and, as a result, was strongly supported. This happened with the new mining policies, especially when the Sección de Fomento (Development Section) was created within the Secretaría de Relaciones Interiores y Exteriores (Ministry of National and Foreign Affairs). The same can be said about efforts to modernize industry by creating a development bank in 1830.

Technological transformation was also taken into account in public works. In addition, special attention was paid to the compilation of federal and state statistics related to population, health, agriculture, stock raising, public funds, and the geographical survey of the country. The development of communications technology, an issue of great interest for the early national governments, relied on corresponding technical knowledge. For example, two technical commissions were created for designing a project to build a coast-to-coast communications network across the Tehuantepec isthmus.

Perhaps the brightest expression of the Mexican state's intention to encourage science was the Instituto de Ciencias, Literatura y Artes (Institute of Science, Literature, and the Arts), opened on April 2, 1826. The institute had fifty members, among them scientific minds of the stature of Andrés del Río (geology) and José Manuel Cotero (chemistry), and men of scientific culture and training, such as Lucas Alamán, José Espinosa de los Monteros, Juan Wenceslao Barquera, and Andrés Quintana Roo. Also, the institute had correspondents in every state of the republic and in Europe (Humboldt, among others), as well as in the new Latin American republics (especially their founders, such as Bolívar, Santander, and Rivadavia). Heading the honorary members were the president of the republic, Gen. Guadalupe Victoria, and the vice-president, Gen. Nicolás Bravo. The institute's bylaws were approved by Congress, and the government backed it "all the way."[2]

Andrés Quintana Roo stated in the inaugural speech that the institute intended "not to teach or to profess a specific art or science, but to foster the advancement and perfection of them all by creating a body that is made up of persons of outstanding ability who, by sharing with each other their knowledge and all kinds of discoveries, can make the taste for education a general trait of the people." In addition, the primary philosophy of this institution would be the same as the state's, that is, "to protect and spread Enlightenment." The 1824 Constitution included authorization to create similar institutes in the other states, and in several of them such institutions began operation immediately.

The application of science and technology was used as policy, and as a means of transforming society, and as a way of shaping the new citizen from the beginning of the newborn republics. Thus, science and technology had a new beginning in Latin America. Science left behind not only its private features but also its eighteenth-century Encyclopedist traits to become an almost bureaucratic discipline closely linked to political interests. This stage and some that followed were marked by the states' indispensable sponsorship and fostering of scientific and technical activities as a result of the political philosophy that inspired the new nations and the absence of competing sectors such as industry and agriculture. In fact, such initiatives belonged exclusively to the state for a long time.

The difficult economic and social circumstances inherited from colonial times along with the lack of social stability after independence and the weakness of the newborn states prevented all anticipated scientific and technological activities from being fully completed at that time. By the third decade of the century, some of the scientific projects planned by the independent Latin American nations had either failed or stopped. In spite of this, liberty and the republican regime had proven capable of creating a nourishing environment for science. Never before had such enthusiasm for scientific and technical knowledge been aroused, or science and technology–based development programs been conceived, as during the early years of independence in these countries. Science and technology showed their potential and what the state could expect from them, as evident in the political will expressed in the different constitutions and in the numerous initiatives for developing and applying scientific and technical knowledge. But the economic reality and the almost permanent state of civil war placed limits on this first stage.

In any case, at the beginning of Latin American independence, a new bond between science and policy, between knowledge and power, was created, a bond that was to be long lasting and crucial for both spheres. A two-sided historical process began: the "scientification" of society, on the one hand, and the politicization of science, on the other. By the mid-nineteenth century, the political triumph of the liberals and, in some countries, their military triumph as well, led to the concept of the utopia of progress. Within this context, science and technology in Latin America were reborn, as the continent's leaders learned from past experience and gave substance to the objectives of the independence generation.

Notes

1. "Francisco José de Caldas to Santiago Arroyo," p. 64.
2. *Memorias del Instituto de Ciencias, Literatura y Artes*, p. 5.

Bibliography

Arciniegas, G. *Bolívar y la revolución*. Bogotá: Planeta, 1984.
Arias-Schreiber, J. *Los médicos en la independencia del Perú*. Lima: Universitaria, 1971.
Arends, T. *Ciencia y tecnología en la época de Bolívar*. Caracas: Fondo Editorial Acta Científica Venezolana, 1986.
Bateman, A. D. *Francisco José de Caldas: El hombre y el sabio*. Cali: Banco Popular, 1978.
Babini, J. *La evolución del pensamiento científico en Argentina*. Buenos Aires: La Fragua, 1954.
Busnell, D., and N. Macauly. *El nacimiento de los países latinoamericanos*. Madrid: Nerea, 1989.
Cáceres, H. *Jorge Tadeo Lozano, vida, obra, época*. Bogotá: Universidad de Bogotá, 1987.
Condarco, R. *Historia del saber y la ciencia en Bolivia*. La Paz: Academia Nacional de Ciencias, 1978.
"Francisco José de Caldas to Santiago Arroyo, April 5, 1801." In *Cartas de Caldas*. Bogotá: Academia Colombiana de Ciencias Exactas, Físicas y Naturales, 1978.
Memorias del Instituto de Ciencias, Literatura y Artes. Mexico City: Imprenta del Supremo Gobierno, 1826.
Rodríguez, L. "El Instituto de Ciencias, Literatura y Artes de la Ciudad de México en 1826." In *Memorias del Primer Congreso Mexicano de la Ciencia y de la Tecnología*, vol. 1. Mexico City: Sociedad Mexicana de Historia de la Ciencia y de la Tecnología, 1989.
―――. "La geografía en México independiente, 1824–1835: El Instituto de Geografía y Estadística." In *Mundialización de la ciencia y cultura nacional*, A. Lafuente, A. Elena, and M. L. Ortega (eds.), pp. 429–438. Madrid: Doce Calles, 1993.
Saldaña, J. J. "La ciencia y el Leviatán mexicano." *Actas de la Sociedad Mexicana de Historia de la Ciencia y la Tecnología* 1 (1989): 37–52.
―――. "Science et pouvoir au XIXe siècle: La France et le Mexique en perspective." In *Science et empires*, P. Petitjean et al. (eds.). Dordrecht: Kluwer Academic Publishers, 1992.

CHAPTER 6

Scientific Medicine and Public Health in Nineteenth-century Latin America

EMILIO QUEVEDO AND FRANCISCO GUTIÉRREZ

Background: Developments in Nineteenth-century European Medicine

Medical concepts and clinical attitudes characteristic of the Enlightenment—called "proto-clinical" by Michel Foucault—changed significantly in postrevolutionary France during the late eighteenth century and the beginning of the nineteenth.[1] Inside the modernizing trends of Renaissance and baroque medicine, such as iatrochemistry, iatromechanics, and vitalism, and nosologic botany-oriented systems,[2] the medical system proposed by Hermann Boerhaave stands out for its hegemonic position.

This system, nurtured as it was by earlier medical developments, integrated elements from the basic sciences of the time (anatomy, modern physics, and a chemistry free from iatrochemical interpretations) with pathology based on clinical observation, just as Thomas Sydenham had proposed several years earlier. Boerhaave founded in Leyden one of the most prestigious medical schools in Enlightenment Europe, and from there his disciples spread his theories throughout the continent.[3]

In spite of this, the multitude of classifications of illness and an inability to completely overcome Hippocratic and Galenic humoral pathology continued to characterize Enlightenment medicine, which maintained the elitist and aristocratic character inherited from medieval times. New political, social, and economic structures brought about by the French Revolution, however, created the conditions for a decisive rupture within the medical tradition of the ancien régime.[4] It closed the hospitals and medical schools, which were considered the ramparts of the old model.[5] In March 1791, in the name of individual liberty, the Legislature decreed that everyone was to be allowed the free exercise of any trade, including medicine. On August 18, 1792, that same body did away with universities and other academic centers.[6] This created a crisis in public health.[7]

This crisis ended in 1794, when, on a totally different basis, new medical schools and hospitals were opened.[8] A new course was embarked on that imposed practical instruction and turned hospitals into the center of medical life; teaching simultaneous with medical attention in hospitals overturned the almost artisanal, dominant forms typical of instruction in the medical schools of the ancien régime.

The Anatomical and Clinical Mentality

In the context of these changes there arose a radical program of conceptual renovation: the anatomical and clinical mentality. It was proposed by Marie François Xavier Bichat in *Traité des membranes* (Treatise on Membranes, 1800),[9] which he included one year later in *Anatomie générale*, where he presented the anatomical lesion as basic to pathology and clinical medicine. In *Anatomie générale*, he combines two processes. First is the clinical approach, based on the observation, description, and systematic classification of symptoms and derived from the work of Sydenham, Boerhaave, and their students and followed by the works of physicians in pre- and postrevolutionary Paris's school of medicine.[10] Second is the great work of Giovanni Battista Morgagni, which established the correlation between the typical clinical symptoms of an illness and lesions in the organs (anatomical pathology) and had great influence in eighteenth-century Europe, especially in Philippe Pinel's *Nosographie philosophique ou méthode de l'analyse appliquée à la médecine* (Philosophical Nosography, or Analytical Method Applied to Medicine, 1798), which states that illness is only a change in tissue or organs and that it is necessary to study them by connecting the symptoms and external signs of illness to suffering and unhealthful changes.[11] Bichat's followers and disciples developed this program through two divergent orientations, which eventually opposed one another: Broussais' *médecine physiologique* (physiological medicine), and the aforementioned anatomical and clinical pathology.

François-Joseph-Victor Broussais was influenced by Pinel and Bichat in the École de Santé (School of Health) in Paris. He was also in contact with followers of Scotsman John Brown, who was a student of William Cullen's. Brown had developed a medical system that stated that "excitability" was the cause of life and of illness.

After his return to Paris, Broussais wrote his scientific manifesto: the *Examen de la doctrine médicale généralmente adoptée* (Study of Generally Accepted

Medical Practice, 1816). In it, he denounces all previous medical customs, attempts to debunk Pinel's nosologic ontology, and exposes the foundations of the latter's "physiological medicine."[12] According to López Piñero, "Broussais thought that life was determined and maintained by the 'irritation' that external stimuli caused in an organism, mainly through the respiratory tract and the digestive tract. Health would be subject to moderate 'irritation.' The excess and—in theory—lack thereof, would cause illness."[13]

Broussais' work was remarkably influential in Europe and in some parts of Latin America; however, in 1830, it began to be supplanted. His opponents favored a more critical and rigorous anatomical and clinical approach.[14] An anatomical and clinical mentality arose, as we have said, as another, simultaneous, branch of the program proposed by Bichat, and it was developed especially by Jean Nicolas Corvisart, Gaspard-Laurent Bayle, and René-Théophile-Hyacinthe Laennec. For them, illness was a lesion located in the body (organ or tissue), and it was this position that defined illness. Therefore, diagnosis was based on pathological anatomy (a discipline developed from systematic observation of thousands of cadavers and the understanding of illness as an anatomical lesion) and semiology (the discipline that studies symptoms and signs and understands the latter as physical manifestations of the anatomical lesion itself and leads us to it).[15]

The philosophical and methodological foundations of diagnosis continued to be those of philosophical empiricism, reassessed from the viewpoint of Condillaquian [i.e., following Étienne Bonnot de Condillac] sensationism. These lie at the basis of the clinical method.[16] All of this implies that close watch over a sick person, the diagnosis, and the later confirmation of the diagnosis during an autopsy require a hospital. The new medicalized hospital is a social invention, thus the term "hospital medicine," which medical historian Erwin Ackerknecht uses to refer to this period of the history of medicine.[17]

Paris thus became, during the first half of the nineteenth century, the unquestioned center of the new science of illness and of anatomical and clinical medicine,[18] and these are the theoretical and practical elements Latin American doctors traveling abroad to be educated in France in the early nineteenth century found.

The Development of Laboratory-based Medicine in Nineteenth-century Europe

By the second half of the nineteenth century, France was losing its central position in European pathological science. The French universities were not able to safely promote the institutionalization of experimental investigation by confining it to a few institutions like the Collège de France and the Musée d'Histoire Naturelle (Museum of Natural History) and separating it from clinical activity. In Germany, on the other hand, scientific-natural pathology was widely developed and superseded the speculative medical systems. It was based on the *Naturphilosophie* (philosophy of nature) of Friedrich Wilhelm Joseph von Schelling and Goethe, which had flourished during the nation's romantic age. According to López Piñero,

> The program continued that transformed pathology into a rigorous science launched from the anatomical and clinical school of Paris. Most important was the overcoming of the antitheoretical posture of anatomists and clinicalists, which reduced pathology to mere clinical and injury observations, totally disconnected from the basic sciences, with its back turned to experimental investigation; a systematic approach was applied to physical, chemical, and biological knowledge, laboratory research becoming the main source of medical knowledge.[19]

The most influential factor in this picture was the unusual setting and cultivation of science and research in Germany between 1825 and 1900, starting with the educational reform carried out by Humboldt. In the new German university, research became a requirement for a career in education, and the "university institute" was the official expression of this process.[20] In the specific area of pathology and clinical science, this organization allowed the doctors from different medical schools to combine their support and educational activities with research.[21]

At the same time in France, centralization and isolation of research related to clinical science resulted in the four disparate roads taken by pathology in this period. The first is represented by Claude Bernard and a small group of investigators,[22] who together leaned on Comtian statements to develop laboratory research,[23] relatively isolated from clinical science.

The second trend, located at the opposite pole, was a continuation of the antisystematic empiricism of the anatomical and clinical schools, and was represented by Ernest-Charles Lasègue, the most famous disciple of anatomist and clinicalist Armand Trousseau.

The third trend, whose greatest proponent was Sigismond Jaccoud, was dedicated to introducing into France the new advances in German physiopathology. This group enlisted clinical scientists with great influence in Latin American medical schools in the early twentieth century, such as Paul Louis Duroziez, Georges Dieulafoy, and Jacques Bouchard, among others. They developed an integrative clinic of ideas taken from diverse schools.

The last tendency was spearheaded by Jean-Martin Charcot. It incorporated laboratory research to bring anatomical and clinical pathology up to date.[24]

In spite of the importance of these groups' theoretical and methodological developments, the laboratory was, in nineteenth-century France, relatively disconnected from hospitals and autopsy rooms.[25] Nevertheless, starting in the late nineteenth century, a positivist conception of illness and practice was established.[26] Thus, the methodology and program that Laín Entralgo calls physiopathological mentality were set up.[27]

Conversely, the same philosophical assumptions and knowledge contributed by the new fields of microbiology and toxicology, the works of Pasteur, Robert Koch, and their students, produced, at the end of the nineteenth century, the idea that illness was the result of the organism's penetration by external agents. This idea was certainly not new: from the Renaissance, beginning with Girolamo Fracastoro and continuing with William Harvey, Athanasius Kircher, Rivinus, Lange, Alfred Hauptmann, and so on, the *contagium animatum* (living organism) hypothesis was enunciated various times. However, the etiopathological mentality built on the foundations of systematic observation in a laboratory was finally brought about by nineteenth-century medicine. Among the pioneers of this trend were Enrico Acerbi and Agostino Bassi, who claimed that contagious infections leading to illness were caused by organized substances able to reproduce, as are all living things. This thesis was supported and further developed by Jakob Henle, who, in *Von den Miasmen und von den miasmatischen Krankheiten* (1840), describes the microbial origin of contagious diseases.[28]

This contribution, which culminated in the works of Pasteur and Koch, had a direct and decisive influence on the way we understand private and public health. From the old Sydenhamian views of hygiene, defined in the private sphere by urban customs and in the public by the effect of environmentally produced miasmas and, to a certain extent, divine will, we moved to a theory of public health sustained by single-cause epidemiology. New theories of public health aimed at eradicating illness—by means of eliminating microorganisms. This view of disease and how it was treated would be very important for Latin American medicine by the turn of the twentieth century.

Background: Medicine in Latin America at the End of the Eighteenth Century

The arrival of modern European medicine, especially Boerhaavian, to colonial Latin America began early in the eighteenth century in some regions, for example, the Viceroyalty of New Spain. However, in its definitive institutional form, this process was directly bound to health reforms begun by the Enlightenment governments of Spain and Portugal beginning in the second half of the century. Starting in the sixteenth century until the crowning of Enlightenment leaders, these kingdoms, as much as their colonies, had remained isolated with regard to scientific advances, medicine, and modern surgery.

The reception, adaptation, and domestication of modern practices in each Latin American region were different because of dynamic, permanent negotiations between strategies and dominant values in the international environment of scientific disciplines, earlier regional cultural developments, and the group and national interests of local intellectual actors and their customers.[29]

From Spain's and Portugal's metropolises came the rationalization of colonial administration to allow these empires to compete with the others in the European market, especially with England and the Netherlands. These Enlightenment governments introduced substantial politico-economic reforms into the structure of colonial exploitation. To bolster them, they felt it necessary also to implement educational and health reforms, primarily to incorporate whites and mulattoes into the colonial project through their enlightenment,[30] but also to improve their productive capacity by raising their life expectancy.[31] Although political intent was common in of all Latin America, educational and health reforms were not identical everywhere.

The proposals coming from Spain and Portugal had to adapt to local conditions; such adaptation did not take place in a mechanical or linear way (for example, in terms of great local development facilitating assimilation of the modern) but in a conflictive and polyvalent process. The end result was an absolutely specific panorama, from which nineteenth-century Latin American medicine would have to be built.

Medicine in New Spain at the End of the Eighteenth Century

New Spain was the crucible in which the home country rehearsed all Enlightenment reforms before trying them out in the rest of the colonies. During the period of Austrian domination, this region was known for the important number of physicians, thanks to significant development of the university and the Facultad de Medicina (School of Medicine) and a strong Aristotelian-Galenic medical culture, based on medical licensing and medical classes.[32] During the Bourbon period, there were attempts to establish lay institutions of learning, as there were in Spain, to overcome the opposition to these reforms in the field of medicine by the university and the medical licensing board (the Real Tribunal del Protomedicato). Health-related institutions included the Real Jardín Botánico (Royal Botanical Garden) and the Real Escuela de Cirugía de México (Royal School of Surgery of Mexico).

After a long process of consultation, the Real Escuela de Cirugía de México was founded on March 16, 1768, by royal decree from Carlos III,[33] following a recent trend of establishing schools of surgery in Spain.[34] Another linchpin in the Bourbon health reforms in Mexico was the Real Jardín Botánico, which played a very important role in the formation of a modern scientific community alongside the university. These institutions, except for the Real Seminario de Minería (Royal Mining College), were imposed by the Crown in the colonies to assure local conditions favorable to the development of the mercantilist economy.

But these health reforms suffered serious setbacks in Mexico. If in the Audiencia of Quito there was fear of the reforms, the state's poverty, the absence of a charismatic figure involved with teaching,[35] and the lack of students all barred the way for progress in Mexico, where health reform arrived early and in all its splendor; the opposition on the part of the university, the Protomedicato, and physicians was authoritative and stemmed from the hegemony of the Scholastic paradigm and from a nascent feeling of nationalism.[36]

The Facultad de Medicina and its graduates,[37] as well as the Real Tribunal del Protomedicato,[38] opposed the Real Escuela de Cirugía on the grounds of a deeply entrenched Galenism and Aristotelianism, and a medieval conception of surgical practice that ignored the progress of this profession in Europe. Something similar happened with the Jardín Botánico, where Linnaean botanical science was attacked by defending indigenous botany.

Changes in the Tribunal de Protomedicato (which comprised professors) were also sought; this led to serious confrontations and hints of rebellion, since change implied a considerable loss of power for the board, for physicians, and for the university. It also represented another form of confrontation between colonists and peninsulars and a defense of autochthonous cultural patrimony.

It is obvious that different ideas about practices and knowledge, guild interests, and nationalist feelings were present in these conflicts. Thus, despite the Crown's interest in promoting change in Mexico and despite the existence of important Enlightenment figures and institutions,[39] the university could not be reformed.[40]

Not until 1833 did the liberal reforms of the war for independence close the university, create the Establecimiento de Ciencias Médicas (Medical Sciences Institution), and introduce the teaching of new French anatomical and clinical medicine.[41]

Medicine in the Kingdom of New Granada at the End of the Eighteenth Century

In the Kingdom of New Granada, including the Audiencia of Quito, change arose more from private initiative than from Spanish imposition. The kingdom's low productivity and demographic density engendered little Bourbon interest in pushing for educational and health reforms in these places. This does not mean there was no spread of knowledge, or, indeed, no exercise of modern scientific practices parallel to activity in the university. Not all innovation took place in institutions, nor did it have to be legitimated by the home country or viceregal power.[42]

Beginning in 1761, with José Celestino Mutis, modern medical and surgical ideas and concepts began to appear. Mutis was educated in the Colegio de Cirugía de Cádiz (College of Surgery of Cádiz) and had wide contact with the intellectuals of the Bourbon court in Madrid from 1757 to 1760, where he assimilated the ideas that prevailed in the minds of Enlightenment Spaniards. He would be the one in charge of proposing and developing reforms in health and medical education.

New Granadan health reform would come very late compared with Spanish and Mexican reforms. It was, however, the most radical in the field of medical education, as much for its content as for its proposed structure. It also unified medical and surgical reforms. Changes in pharmacy were not as clear; however, a botanical expedition did take place, but from a medical point of view, it did not play as important a role as the botanical gardens of Madrid and Mexico.

Unlike in New Spain, in Santa Fe and Lima, pedagogical reforms, in spite of the delay in implementing them, were much more stable because of the absence of a strong Aristotelian-Galenic medical culture and the lack of a cohesive, highly institutionalized, pre-Columbian medical tradition. The influence of the Bourbon educational reforms in the health field persisted until the first half of the republican century.

It is important to point out here that Spanish politics concerning the Crown's overseas territories was selective and unequal, among other reasons, because of internal and external difficulties caused by conflicts and rivalries with other European powers that prevented Spain from throwing its net over everything. On the other hand, the risks naturally related to technical development and a framework of social realities also played a remarkable role in defining the final orientation and achievements of Bourbon scientific and health reforms.

Medicine in Portugal and Brazil at the End of the Eighteenth Century

The Enlightenment had also been felt in the Lusitanian region. The reforms enacted by the Marquis of Pombal, one of José I's ministers during the 1750–1777 period, were guided by a spirit of enlightened despotism and modified the economy, administration, health, and education. In this last field, there was a trend toward laicization and secularization, and to this end, the Jesuits were expelled from the Lusitanian kingdoms.[43] In the eighteenth century, the medical school at Coimbra was still the only one in the Portuguese world. Pombal initiated changes in 1772,[44] in a spirit similar to that of the Spaniards.

In Brazil, there was no medical education until the end of the eighteenth century. Doctors came from Coimbra, surgeons from the Hospital Real (Royal Hospital) of San José de Lisboa. It was not until 1803 that São Paulo's captain general, Antonio José de França y Horta, officially instituted a surgery class in the military hospital of the Paulist capital, though it had only six students.[45]

It would not be until Dom João transferred the seat of the Portuguese state to Rio de Janeiro that the regular teaching of medicine and surgery in Brazil would be considered justifiable. Surgery then became a priority.[46]

The Transition: From Colonial to Anatomical and Clinical Medicine in Latin America

In general, the colonial state considered health to be a private matter, something that people should take care of themselves. The state needed only to assure control of medical practice—by means of the Protomedicato—by guaranteeing that medical professionals were properly trained. The population's good health would then be the result of enough good physicians. For this reason, the state concerned itself with establishing medical schools where there were significant numbers of persons of European origin and used medical licensing boards to assure control and efficiency. In Spain, municipalities had an employee, called a *mustasaf*,[47] who took charge of hospitals, food, and public hygiene in general;[48] however, this was more an issue of politics and sanitation engineering than of medicine.

By the end of the eighteenth century, some medical ideas related to the changes in postrevolutionary medicine in France had come to Latin America. Latin American medical philosophy was thereby undergoing a slow transition to anatomical and clinical concepts. This phenomenon was reinforced and stimulated by French cultural and political processes of national independence in the different Latin American countries. The political attitude of these new states with regard to health, however, stagnated for a while. To study this process, we will look at three basic aspects: medical schools; health-related policy; and health research.

The Transition to Anatomical and Clinical Medicine in New Granada and Gran Colombia

The Facultad de Medicina in the Kingdom of New Granada was reestablished in the Colegio del Rosario in 1802, under the guidance of José Celestino Mutis. During the colonial period, it was nonfunctioning for all intents and purposes: it graduated only two doctors between 1636 and 1765. The school was reopened after an important polemic concerning the problem of health and medicine in New Granada in the twilight of the eighteenth century.[49] The school functioned until the War for Independence began in 1810, to be reopened after independence in 1819. During this period, a new class was inaugurated in the Colegio de San Bartolomé de Bogotá, directed by one of the Colegio del Rosario's disciples, José Félix Merizalde. The program of study included French authors inclined toward anatomical and clinical medicine, but it did not teach these views in any explicit way.[50]

In New Granada after independence, the programs of study from Rosario and San Bartolomé were restarted in 1826, at the Facultad de Medicina of the recently created Universidad Central. Mutisian medicine was the program's main focus; however, the doctrines of Brown and Broussais, recently imported by the English and Irish doctors who came with the military legions sent to support the struggle for independence and by the French who came to work in the Museo Nacional (National Museum) in 1822, were soon introduced.[51]

In the field of public health—in the period immediately following independence—the New Granadan state maintained the same political attitude as Spain: when he created the Universidad Central, Pres. Francisco de Paula Santander awarded responsibilities formerly exclusive to the Real Tribunal del Protomedicato to the Facultad de Medicina beyond normal teaching functions. Following the university's founding in 1826, the continuous turbulence (the breakup of Gran Colombia; two large uprisings that drove New Granada to the edge of dissolution, in 1832 and 1840, respectively; multiple local rebellions; territorial secessions, etc.) had a direct effect on educational politics and led to a gradual weakening of all institutions, including those around which the medical communities were grouped. This ended with the dissolution of all universities in 1850.

In this environment, physicians were in constant conflict with various traditions (the ideas of Brown, Broussais, and Mutis), which somehow led to a catastrophic balance. These three approaches to understanding illness and medical practice remained mutually opposed until the promulgation of the Law of May 15, 1850, which closed the universities; declared the teaching and exercise of all professions, except for pharmacy, to be free; and eliminated titles as prerequisites for professionals.[52]

New Granadan medicine was immersed in a crisis from which it would never emerge: New Granada became the Estados Unidos de Colombia (after multiple civil wars that culminated in the Constitution of Rionegro, 1863), and the medicine practiced in the first half of the nineteenth century—together with its international sources (Boerhaavianism, Brownism, and Broussais' physiological medicine)—would forever disappear from the country.

For more than a decade, the disintegration of academic medicine was addressed by a blossoming of empirical medical practices that could neither satisfy the needs of public health nor meet the demands of the elite. The elite became dissatisfied by the predominant "anarchy." As a consequence, some of their children, facing the breakup of institutions and medical traditions in the country, turned to France,[53] the main point of cultural reference.

This handful of Colombian pioneers had a chance to be dazzled by already solid anatomical and clinical practice in Paris's classrooms and to witness the first steps of physiopathological and etiopathogenic ways of thinking. Once back in Colombia, they reestablished the Facultad de Medicina on new conceptual and organizational bases. This movement led to the development of three important phenomena that strengthened the rupture with previous medical schemes.

In the first place, to separate themselves from empirical physicians who could now practice freely, Colombia's new generation of doctors began to form associations in an effort to maintain their professional identity and to defend the profession. They also acknowledged the need to study and understand national pathology and to build medical standards for the country.

In the second place, and as a manifestation of the same phenomenon, they promoted the publication of the first scientific publications. *La Lanceta* (The Lancet),[54] the first medical journal in Colombia, published its first issue in 1852; *La Gaceta Médica de Colombia* (The Medical Gazette) appeared in 1864.[55] Their goals were, according to Néstor Miranda, "to serve as a way for Colombian physicians to share information and experiences; to allow the spread in Colombia of advances in international medicine; to be a vehicle for solidarity of the 'body of physicians' in terms of both science and unity, and to allow the development of group solidarity . . .; to fill the role of spokesman and pressure group before the state and civil society to force both to recognize the medical profession and to bar the way 'to empiricism and quackery.'"[56] These publications—mainly the *Gaceta*—had an anatomical and clinical orientation. Thus, the *Gaceta* became an extremely effective weapon in the fight to reestablish the medical profession and its medical institution, in particular, in the academic realm.

Finally, the Escuela de Medicina grew in the private arena and, although it was short-lived, it continued in the form of the Facultad de Medicina of the Universidad Nacional de Colombia. Trained in Paris in anatomical and clinical thinking, Dr. Antonio Vargas Reyes, intellectual soul of the *Lanceta* and the *Gaceta*, had a complete understanding of what hospitals represented in the formation of physicians. Thus he promoted, from the *Gaceta*, the creation of a private medical school that was tied to the hospital.[57] Editorials were published in several issues calling for the founding of the Escuela de Medicina, which happened relatively quickly. It opened in January 1865, and after six months was a success, with an openly anatomical and clinical program of study. From this date, the journal took the subtitle "Organ of the Escuela de Medicina."

In the area of public health, things did not change substantially. In the political arena, the state in general did not modify its attitude; in the academic area, the teaching of anatomy and clinical medicine occupied hardly any space in the medical curriculum, remaining reduced to a hygiene class.

The Transition in Mexico

In Mexico, the Enlightenment was developed around a handful of strong personalities and parauniversity institutions unconnected with classic universities. The latter, in general, kept to the Scholastic tradition. Thus, it was relatively easy for the independence movement—certainly quite disputatious[58]—to deal with an institution with neither progressive nor national bonds. Parauniversity institutions (the Real Escuela de Cirugía de México, the Real Jardín Botánico, and the Real Seminario de Minería) in which the Enlightenment had entrenched itself quickly decayed. The independence movement, while physician Valentín Gómez Farías was the republic's vice-president, closed the Real y Pontificia Universidad de México (Royal and Pontifical University). The Dirección General de Instrucción Pública (General Office of Public Instruction) was created to reorganize education. Regarding medicine, the former Facultad de Medicina of the university together with the Real Escuela de Cirugía became the Establecimiento de Ciencias Médicas (Medical Sciences Institute) in 1833. On October 23, the program of study of this institution was published, and Dr. Casimiro Liceaga was appointed director.[59] The great majority of its professors were graduates of the Real Escuela de Cirugía, a group of Enlightenment surgeons who sought to unite practical learning and the medical theories then in fashion.[60]

Eight months later, the reform of the university was canceled, the university was reopened, and the Dirección General de Instrucción Pública was closed. But the Establecimiento de Ciencias Médicas received special treatment and remained active as the renamed Colegio de Medicina.

There some professionals welcomed the ideas of Brown and Broussais, but the most critical, for example, Manuel Eulogio Carpio, quickly veered toward anatomical and clinical views.[61] In the new academic institution, anatomical pathology and physiology were taught. We do not know, however, whether the anatomical pathology taught there was Bichatian or not. We do know that the physiology class taught by Dr. Carpio from the beginning followed the experimental trend that François Magendie had been working on in Paris;[62] this transition, in contrast with the very painful one in New Granada, occurred almost without conflict, despite academia's economic and institutional uncertainty.

The Transfer of Anatomical and Clinical Medicine to Brazil

In Brazil, the absence of medical teaching institutions during the colonial period was absolute. The need for support in the health area was not felt until 1808, when Dom João and the Portuguese court moved to Brazil and established a metropolis in the periphery. Schools of surgery were founded in Bahia and Rio de Janeiro,[63] cities with the largest white population. These schools were inspired by the spirit of modern surgery but adapted to Brazil's unreliable supply of surgeons and physicians.[64]

The creation of the two medical-surgical courses was linked to the reorganization of health services. The regent reestablished two positions that had not existed since 1782: primary physicist of the kingdom and surgeon major of the army, transforming them into the first health authorities of Portugal's administrative organization. Together they constituted a kind of general inspectorate of public health. The Escola da Cirugia e Anatomia (School of Surgery and Anatomy) was created at the suggestion of the king's surgeon major, José Correia Picanço. The school in Rio, created soon after, was justified by the need of the Hospital Militar y de la Marina (Military and Navy Hospital) for surgeons who could treat illness contracted on ships and by people coming to the hospital.[65]

The creation of these courses, however, did little to modify the training routine in hospitals, which maintained the same empirical system of teaching that was used before they were established. As a consequence, in 1811, the government charged Dr. Vicente Navarro de Andrade, a physician recently arrived from the university in Coimbra, with the task of revising medical and surgical courses. He proposed an integrated school in which medicine, surgery, and pharmacy would be taught together, using state-of-the-art anatomical, clinical, surgical, and obstetric techniques. This program of study was never put into practice, however, and it was not until 1813 that the school in Rio de Janeiro was reorganized, although less ambitious but better adapted to local conditions, according to a plan proposed by Manuel Luis Álvares de Carvalho and its name changed to the Academia Médica-Cirúgica (Medical-Surgical School).[66] The school in Bahia was reorganized in 1815, following the pattern of the school in Rio.[67] Enlightenment surgery entered Brazil and adapted to local conditions without any trauma.

The proclamation of independence in 1822 did not mean immediate modifications in medical-surgical teaching or health services. But the classic rivalry between physicians and surgeons was not long in coming. As a result of the increase in social recognition of new French medicine—as an activity implying knowledge beyond typical old-fashioned knowledge shared by surgeons, leeches, midwives, and so on—in 1829, the Sociedade da Medicina (Medical Society) was founded on the model of the French Académie des Sciences. Five prestigious physicians were involved in its founding—two Brazilians and three foreigners—as were two teaching surgeons, their purpose being to improve the status of the profession and to differentiate themselves from working surgeons.

Their first assignment was to study suggestions for the revision of medical teaching. A commission was named and, after a year, it presented a proposal that was transformed into the Law of October 3, 1832. This law, in turn, transformed medical-surgical academies in Bahia and Rio into schools of medicine, with the right to award doctorates in medicine, pharmacy, and midwifery, but abolished the title of bloodletter.[68]

From a curriculum viewpoint, the "void" in medical academic traditions allowed medical schools to import, without resistance, new auxiliary sciences, such as physics, chemistry, and botany, as well as the firmly established French anatomical and clinical science. From the social point of view, the insertion of home country elites into the periphery allowed doctors to, on the one hand, work in politically stable conditions and, on the other, be able to count on a sure clientele that could provide social status and appropriate economic recompense.

Consolidation: From Hospital-based to Laboratory-based Medicine

Once the possibility of anatomically and clinically oriented teaching was established, although in different times and under different circumstances in each country, a paradoxical consolidation process began as well. In Colombia, as a result of the efforts by a circle of doctors trained in France, the pattern was one of strong clinical science, supported by anatomical pathology, some later contact with etiopathology, and a slight link with physiopathology.

This did not occur in Mexico until the beginning of the 1920s, under the impetus of an important nucleus of locally trained Creole doctors. Anatomical/clinical and physiopathological trends consolidated quickly and firmly, and there was balanced development of theory in these areas.

Although development was colored by institutional uncertainty in Brazil, the mixture of metropolitan and Creole doctors, the country's geographical and climatic conditions (favorable for the growth of tropical disease), and the absence of an academic medical tradition supported by the state favored the acceptance of microbial theory and the development of important research in the fields of microbiology and epidemiology. This research led to an active public-health program. By the end of the nineteenth century and the beginning of the twentieth in Europe and Latin America, in the process of public-health organization, the development of bacteriology—and the concept of etiopathological illness derived from it—played an important role in the final crisis of all nineteenth-century medical theories.

The Development of Anatomical and Clinical Medicine, Physiopathology, and Microbial Theory in Colombia

In Colombia during the second half of the nineteenth century (1865 forward), local doctors led by those trained in France and proponents of the anatomical and clinical approach campaigned for the rejection and replacement of common medical ideas and practices. Their approach was framed by the cycle of federalist liberalism, persisted throughout the conservative tenure, and ended with the beginning of financial dependence on the United States well into the twentieth century. Their leadership was characterized by a growing nationalist concern,[69] an attempt to research and study local pathology, and an inclination toward establishing a national medical corps.[70]

This period can be divided in two stages. During the first, hospital-based medicine took hold and comfortably settled in. The beginning of this process was clearly marked by the founding of the Escuela de Medicina, mentioned above, in 1865. The second phase is notable for the slow progress of laboratory-based medicine, especially etiopathological procedures. Its final stage, starting in the 1950s, is defined by the introduction of Flexnerian reforms from North American technological medicine.

The boundaries between the first and the second stages, however, are not clearly defined, since, unlike what happened in Europe, physiopathological and etiopathological ideas did not break into programs of study. There was instead a slow adaptation to these ideas and a "negotiation" that allowed their adaptation to the hegemonic conceptual structure of anatomical and clinical science.[71]

In the field of medical education, after clerical modifications by conservative governments following radical liberalism and the revolution of 1860, a new interest in university education began. It resulted in the founding of the Universidad Nacional in 1867, and its medical school.[72]

The medical personnel involved in the *Gaceta* found itself at a crossroads: they had to decide whether to support the privatization project represented by the Escuela de Medicina or to participate in the state's proposed Universidad Nacional. Although one might think that pure self-interest and a spirit of close unity would lead the medical community to rally around the privatization project, the exact opposite happened. The Escuela de Medicina, after only two years, was voluntarily closed by its professors, who moved, bag and baggage, to the Universidad Nacional.

There are several reasons for this. On the one hand, France, acting as a collective imaginary fortress, was an obligatory and paradigmatic reference for the elite in general and for the medical community in particular; in France, medical instruction was the responsibility of the state. On the other hand, the "idea of progress" was a theme for members of all the political parties. If we tie this to the doctors' deeply felt need to adapt their modern knowledge to the understanding of disease in the periphery—a need that "naturally" took the guise of a civilizing, Promethean action—we find once more the bridges joining the elite's goals with those of the medical community. Thus, doctors felt that in some way their civilizing profession was in harmony with the Colombian state's proposals.

Finally, the existence of diverse traditions all moving toward unifying physicians with the state should not be overlooked: the prototype of the professional public servant (so criticized by the press of the day), and the existence of a former medical school linked to the state university, the Universidad Central.

The new academic program of the Facultad de Medicina of the Universidad Nacional was almost exclusively anatomical and clinical. Beginning in 1886, its regulations were modified to remarkably resemble the 1878–1883 program of study of the medical school in Paris. The French institution's influence was very important throughout the period, not just in Colombia but in much of Latin America as well. Buenos Aires's Facultad de Medicina, for example, in 1880 approved an academic program that was a copy of the 1878 Paris program.[73]

In the city of Medellín, capital of Antioquia, there were ill-fated attempts beginning in 1834 to create a medical class; the constant civil wars, among other things, worked against educational continuity. In October 1871, the creation of a school of medicine in Antioquia was proposed,[74] but did not actually happen until 1872, with the establishment of a fundamentally anatomical and clinical program of study.

Laboratory-based medicine was built around Comte's physiopathological approach and positivism, both of which entered the country much later,[75] because late nineteenth-century intellectuals, the children of a curious mix of deistic Spanish Enlightenment and liberalism, were unconvinced about this philosophy.[76] In the last decade of the nineteenth century, microbial theory was most influential and affected both health-related research and politics, even though, as we have already noted, teaching methods were not modified in any meaningful way.

In Colombia, this is clearly exemplified in a talk given by Dr. Pablo García Medina on July 21, 1897, to the Academia Nacional de Medicina (National School of Medicine). There is almost no mention of the fundamental contribution of Claude Bernard, which consisted of "having established the principles of experimentation applied to the study of living beings, in both the normal and the pathological state," but there is a constant insistence on the importance of Pasteur's microbial theory. While Bernard is given two pages in García's text, microbial theory, and its importance to and applications in the fields of medicine and public health in Latin America, get a total of thirty-five pages.[77]

The physiopathological view required a much more complex infrastructure, and its practical application did not appear as obvious to Colombian physicians as did etiopathology. Until World War I, there was no necessary minimum allowing for the development of laboratory-based medicine in Latin America.

Laboratory-based Medicine in Brazil

The transfer of the court to Brazil caused deeply contradictory local transformations because it implied the transplantation of metropolitan society to a colonial city structure. The economic measures consequently adopted, such as the opening of ports, also meant the transformation of urban and port structure, seat of new colonial–home country commercial relations.

Brazil's cities were not ready for this change, which came with the adaptation of urban space to new political, economic, and social requirements by a nobility and a middle class accustomed to life in metropolitan centers. The task was often hindered by local epidemics. The lack of physicians justified the government's interest in developing the new schools.[78]

The creation of medical schools beginning in 1832 meant the formation of an institutional space that was invaluable to the social reproduction of medical knowledge aimed at, on the one hand, the formation of a "healthy consciousness for the people" and, on the other, the institutional exclusion of quacks and *mandingueiros*, witch doctors who practiced healing magic that deviated from the medical norm. This process of institutionalization, however, met opposition not only from already established Portuguese physicians, who rejected any practitioner not trained at Coimbra, but also from the bulk of the popular classes.[79]

From a curriculum standpoint, the new medical schools became a vehicle for the adoption of metropolitan models of hospital- and laboratory-based medicine. This fact notwithstanding, when we looked at the first theses written by graduates of the Faculdade de Medicina (School of Medicine) of Bahia, we found medical treatises permeated by disparate philosophical trends and by the interests of the dominant social sectors. There was a fierce fight between, on one side, the vitalist and eclectic beliefs common to Enlightenment medicine and, on the other, positivism, which slowly introduced the experimental, empirical method, supported by quantitative science and the development of medical technology.[80]

Despite this, in the nineteenth century, medical research was not an integral part of the common physician's activities. However, the growth in European medical knowledge during the second half of the century sparked a new research phase.[81] According to Brazilian Comtians, the branches of science that would help progress were those with the ability to promote knowledge over reality, the certain and indubitable, the useful and precisely determined. At the Rio and Bahia schools, it was thought that research should be institutionalized in the form of theses required for a degree.[82]

Thus began the development of an intense interest in epidemic disease. The schools and the Sociedade da Medicina (Medical Society) played an important role in research and advised the state during health crises. Many medical magazines were founded, such as the *Propagador das Ciências Médicas* (Propagator of Medical Science), published by Joseph F. X. Sigaud (a founding member of the Sociedade da Medicina) and the Seminário da Saúde Pública (School of Public Health).[83]

During the curriculum reforms of 1854 and 1879, positivists tried ever harder to ensure their own hegemony. Despite this, clinical tradition continued to dominate experimental practices, the lack of laboratory facilities in schools was remarkable, and centralist bureaucratic hindrances did not allow for the betterment of the situation. A change would not be made possible until 1880, when, with the advent of the liberal republic, the government funded some laboratories. It created the Departamento de Microbiologia (Microbiology Department) in 1901 and required attendance at labs.[84]

In the research field, the Escola Tropicalista Baiana (Tropicalist School of Bahia) was an exception.[85] Such was the denomination of a medical movement started in 1850 and developed independently from the Faculdade de Medicina and around the Santa Casa da Misericórdia Hospital and the *Gazeta Médica* of Bahia. It was founded by Drs. John L. Paterson (a Scottish doctor in charge of the health of the British community in Salvador, Bahia), Otto Wucherer (a Portuguese doctor of German descent), Francisco Da Silva Lima (a Brazilian), Thomas Wright Hall (an Englishman), and others. They created the pivotal idea of a "tropical pathology" to study health disorders caused by warm climates and to discredit the idea of the unhealthfulness of tropical zones. They proved that tropical disease could be understood and fought in the same way as diseases of the cooler climates.[86] The Tropicalist school thus developed an original theory and proposed a nonstandard knowledge set, although it was not totally autochthonous, since it was strongly linked to the Anglo-Germanic model of experimental pathology.[87]

An institutionalized research tradition appeared by late 1899, when bubonic plague erupted in Santos, in the State of São Paulo. Few Brazilian doctors were familiar with recent works on this disease. The State of São Paulo had begun, in 1892, a comparatively far-reaching public-health system, and the epidemic was quickly aborted, with important help from Adolfo Lutz, chairman of the Instituto Bacteriológico (Bacteriological Institute), and his disciple Vital Brazil.

Conversely, the plans for the defense of the City of Rio de Janeiro against the plague were rather confused. The prefecture commissioned a young bacteriologist recently returned from three years' study at the Pasteur Institute, Dr. Oswaldo Cruz, to accompany the well-known surgeon Dr. Eduardo Chapot-Prévost to Santos. There they performed autopsies on the victims and confirmed what Lutz had already discovered: the plague came from overseas. First cases presented in Rio around 1900, claiming the lives of 295 people.

Duplicitous, anarchic organization in Rio made it impossible to mount an effective campaign against the plague until Oswaldo Cruz was named director of the Direitoria Geral de Saúde Pública (General Directorate of Public Health) and reorganized it in 1903. The prefecture bought an old hacienda in the remote parts of Manguinhos on which to erect laboratory facilities for the study of the plague bacillus, with the aim of producing a vaccine. Thus the Instituto Federal de Seroterapia (Federal Serotherapy Institute) of Manguinhos was created in 1900.

There were fruitless attempts to hire a bacteriologist from the Pasteur Institute. Prof. Emile Roux, disciple of Pasteur and associate director of the Pasteur Institute, suggested Oswaldo Cruz, who had recently ended his training there. In May 1900, at age twenty-eight, he took over as director of the new Instituto Federal de Seroterapia and begin a career as a researcher of great consequence in the development of microbial theory and laboratory-based medicine in Brazil. The Instituto de Manguinhos, modeling itself after the Pasteur Institute, became, in the twentieth century, the place for medical-bacteriologic research in Brazil; important discoveries and contributions to world medicine have been made there, and it has become the headquarters for important local and national health campaigns.[88]

Physiopathology and Etiopathology in Mexico

The program of studies for the Establecimiento de Ciencias Médicas was revised in 1834 to require the study of certain subjects in the reestablished university. In 1843, experimental physics and medical chemistry were introduced.[89] In 1857, 1862, and 1894, the teaching of clinical science based on French models was incorporated, and the institution was renamed the Escuela Nacional de Medicina (National School of Medicine).[90] Unfortunately, we cannot locate more detailed information about these reforms, but we can deduce from clinical stories by Dr. Miguel Francisco Jiménez, professor at the school and quoted by Martínez Cortés, that a clinical, anatomicophysiological approach was being used, with the physical support of lab data.

Researchers at the Academia de Medicina, beginning in 1879, began to study the etiology of disease. Their work generated competition by promoting studies of timidly etiopathological orientation.[91]

The Museo de Anatomía Patológica (Museum of Pathological Anatomy) was created in 1895, to which chemistry and microscopy sections were added one year later in the service of clinical science. In 1896, this institution began the *Revista Quincenal de Anatomía Patológica y Clínica Médica* (Journal of Pathological and Clinical Anatomy) to publish the research that was developed there. In 1896, it became the Instituto Patológico Nacional (National Institute of Pathology), and still later became the Instituto Bacteriológico Nacional (National Bacteriological Institute), which promoted etiologic and microbiological studies, derived from it. Dr. Eduardo Liceaga introduced Pasteur's ideas into the country and in 1888 founded the Instituto Antirrábico Mexicano (Mexican Antirabies Institute), which studied rabies and where rabies vaccination was headquartered.[92]

From Hygiene to Public Health in Latin America

From the second half of the nineteenth century to the beginning of the twentieth, events and circumstances favoring the structuring of national health organizations took place. First, the development of microbial theory displaced Sydenham's epidemiological, miasmatic theories and the resulting bacteriological research establishments, such as the Pasteur Institute. Second, international health organizations were created, especially the Pan American Health Organization. Finally, in the twentieth century (and outside the scope of our topic), North American philanthropic foundations appeared, specifically, the Rockefeller Foundation.

From Hygiene to Medicalized Public Health

Starting with the discovery of the relationship between disease and microorganisms and the appearance of laboratory-based medicine—with its great advances in the determination of causal agents for disease—whether biological (bacteria, viruses, parasites, etc.), chemical (poisons, toxins, etc.), or physical (radiation, trauma, etc.)—unicausal epidemiology and huge new therapeutic measures arose as effective weapons in the eradication of disease. It was thought that once the causal agent of a disease was known, hygiene alone would determine the factors that favored or hindered its onset and would determine measures to manage or control it. Research institutes began to appear in Europe especially concerned with the production of vaccines to prevent disease, since these, together with eradication campaigns, appeared to be the only weapons for a public-health philosophy that had abandoned the classic concept of hygiene.

This rationale appeared more overwhelming when applied to the common infectious tropical diseases in Latin America. The Pasteur Institute, for example, had great influence on health-related programs in tropical countries in the second half of the nineteenth century.

International Health Organizations

Attempts to create a global health organization started with the first International Sanitary Conference in Paris in 1851. Conferees gathered for the purpose of imposing quarantines in Europe, the main weapon against the spread of disease from one country to another at the time. Because of the multiple difficulties in finding common ground when it came to methods and technical knowledge, a quarantine for cholera was not agreed on until the seventh International Sanitary Conference, in 1892. As a result, the First International Health Convention was drafted.[93] In 1912, the twelfth International Sanitary Conference agreed on a regulation regarding quarantines for cholera and plague but barely considered yellow fever.[94] This last, however, was the most serious problem for Latin America.

By the second half of the nineteenth century, the United States was commercially competitive on the global scale, but the country did not concern itself with matters of hygiene until 1880, when Congress authorized the president to convene the fifth International Sanitary Conference, in Washington, D.C., "for the purpose of organizing an international system of notification about the true sanitation condition of ports and plazas."[95]

This position was the result of an economic and social situation generated by continual cholera outbreaks (caused by European immigration to the North and South American continents) and a progressive increase in cases of yellow fever resulting from commerce with Cuba and Central and South America. By 1879, the U.S. government had tried to impose regulations in the ports with which it traded, but only a new global conference could help it succeed.

In 1893, the U.S. government authorized the chief of the Marine Hospital Service, Dr. Walter Wyman, to deploy medical officers from the service to the consular offices located in main shipping ports. This decision caused a lot of uneasiness in Europe and Latin America, so the second International Sanitary Conference of American Republics was held, in Mexico City in 1901. Conferees concluded the following: "The Administrative Council of the International Union of American Republics [today the Organization of American States] should summon a 'Convention' of representatives for American health administrations, so as to formulate 'agreements and health regulations' able to 'reduce to a minimum the quarantine requirements' regarding cholera, yellow fever, bubonic plague, pox, and other 'grave, pathogenous' outbreaks."[96] This convention established the first of what would later become the Pan American Sanitary Conferences, held every four years. The organizing committee became the International Sanitary Bureau; it would later be renamed the Pan American Health Organization, with seven members, three of them U.S. citizens: the surgeon general of the United States and two of his or her subordinates; the four remaining seats were to be rotated among the other countries of the continent.

It became evident that control of health-related organization in the Americas belonged to the United States; in fact, the surgeon general of the United States was elected to direct the Pan American Health Organization until the 1960s.[97] The second half of the nineteenth century was therefore notable for being a period of transition between European hygiene and a public health culture originating in the United States.

From Hygiene to Public Health in Colombia

In 1886, Colombia enacted a new Constitution, which began what we have called the "hygienist period."[98] This Constitution established a new centralist legal framework for the organization of public health as a state function. Law 30 of 1886 was promulgated, and as a result, the Junta Central de Higiene (Central Hygiene Board) was established as the first official entity charged with solving problems related to public health. This hygienist model, influenced by the United States, culminated in the creation of the Ministerio de Higiene (Hygiene Ministry) by order of Law 27 of 1946 and Decree 25 of 1947.

Let us review. The defeat of federalist tendencies in the Colombian state at the end of the nineteenth century, expressed in the Constitution of 1886, created a new, frankly Catholic, attitude toward the people, specifically, toward their health needs. Its architects understood the social dynamics of the second half of the nineteenth century in Europe and on the Latin American continent. The criticism of romantic liberalism and of laissez-faire and its philosophical foundations convinced them that it was necessary to reinforce traditional institutions—such as the church and the family—as vehicles of social cohesion.

Facing the social conflict arising from incipient industrialization, the state turned to Catholic social doctrine as propounded by Pope Leo XIII. An important part of this doctrine was linked to the protection of the underdog. All of this led to a constitutional document allowing national unification and state intervention in society, the economy, education, public works, health, and so on, with a good dose of authoritarianism and traditionalism.

The contrast between the politics of this new group, which, under the leadership of Pres. Rafael Núñez,[99] favored regeneration, and the romantic radicalism of the federalists is clear in both affairs of state and the attitude toward public health. The most important activities of the Junta Central de Higiene aimed at taking measures to control the best-known epidemic diseases of the day (leprosy, cholera, smallpox, plague, and yellow fever); to improve management of water, garbage, food, and so on; to improve sanitation in public entities (asylums, hospitals, schools, etc.); and perhaps most important, to apply the international quarantine laws to avoid both the entrance of new infectious diseases and isolation from international trade.[100]

The quarantine regulations were only partially enforced during the last two decades of the nineteenth century, since it was not until the early twentieth century that the government legally accepted the regulations proposed by International Sanitary Conventions, with the enactment of Law 17 of August 21, 1908, which approved the Sanitary Convention *ad referendum*, concluded in Washington on October 14, 1905, and signed by Colombian delegates during the third Pan American International Sanitary Congress, held in Mexico City on December 2, 1907.[101]

Although the Junta Central de Higiene performed an important service in the areas outlined above, there was not yet a national health organization; there were no departments or municipal or rural representatives of the organization. Attempts to centralize and unify the nation in 1886—beyond the health field—were opposed by diverse groups for a few decades. Some obstacles were temporary, others were more structural. A national state could not be established on the basis of precapitalist economics or obvious territorial fragmentation and the lack of an intellectual or political elite familiar with modern state mechanics.

Nearly twenty years after the promulgation of the 1886 Constitution, known as the "Regeneration Constitution," the country was still "pastoral." There was ignorance of conditions indispensable to a modern state: industry that incorporated technology; agriculture made over by technology and organically linked to it; mathematical science; defense attentive to technological advances and geographical determinations. All of these led the country to manage international issues in the manner of a precapitalist lord.[102]

Hygiene in Mexico

There is little information about hygiene in Mexico. We can affirm only that from the opening of the Establecimiento de Ciencias Médicas in 1833, hygiene began to be taught from the Sydenhamian perspective of a set of individual physiological norms, to which hygiene was a sort of appendix. In 1868, both subjects were separated, and in 1889, Dr. Luis E. Ruiz modified the teaching of hygiene from a positivist perspective, focusing on the problem of disease prevention from a triple standpoint: general hygiene, in which the human world is studied; special or individual hygiene, which studies human beings and their functions; and public or social hygiene, which covers the control of collective illnesses and health regulations.[103]

The Instituto Antirrábico Mexicano began vaccination against rabies and tuberculosis, and in 1891, the Código Sanitario de la Capital de la República (Capital of the Republic Health Code) and regulations pertaining to beverages and groceries were published.

Hygiene and Public Health in Brazil

Starting in 1826, when medical instruction became independent from the Fisicatura Maior (Board of Medical Examiners), the governmental organization in charge of medical issues inherited from the colonial period, the Fisicatura started to appear inefficient and to decay. It was an odd entity in a newly established nation and became identified as a defender of Portuguese interests in Brazil. In 1828, this old-style example of power disappeared.

The establishment of the Sociedade de Medicina (Medical Society) engendered fresh discussions of medical power, which were not acknowledged by the state until the society was transformed into the Academia Imperial de Medicina (Imperial School of Medicine) six years later. From that moment on, it worked with the state to intervene medically in society on behalf of the state. It focused on declaring civil unrest to be an etiological agent of disease, and thus under the jurisdiction of the medical community, thereby introducing a new "moral order–health" relationship. This idea assured an alliance between doctors and hegemonic social groups, so maintaining health also meant enforcing the institutional social order.[104] This attitude toward health prevailed throughout the imperial period in Brazil.

The organization of health services was extremely precarious and based on miasmatic theory. Until the mid-nineteenth century, hygiene was the responsibility of local authorities, who were mainly concerned with cleaning refuse from streets and homes, purifying water and air, and building latrines. Medical assistance for the poor depended on philanthropic initiatives on the part of socially important figures and church-based charitable institutions, such as the Santa Casa da Misericórdia. The rest of the population contracted existing medical services privately. In 1850, one year after the yellow fever epidemic in Rio de Janeiro, the Junta de Higiene Pública was created to unify the empire's health services.[105] Beginning in 1889, during the liberal republic, health policies were noted for their deference to the new Brazilian liberals and their political and economic interests. Health practices veered toward control of a limited group of illnesses—those that threatened the work force or the expansion of capitalistic economic activities in rural and urban areas.[106]

The 1889–1930 period registered the introduction and establishment of a service mode that used available tools to combat epidemics and that favored studies of different etiologies, precise diagnostic media, and generalized immunization measures. This period also saw two important political events. First, the "Oswaldo Cruz turning point," which marked the introduction to the country of a health organization based on bacteriology, microbiology, and immunization.

This juncture solidified in 1902, when, under Pres. Francisco de Paula Rodrígues Alves—spokesman for the coffee tycoons—various urban and health reforms were proposed to allow immigration, believed at the time to be the basis of economic progress. Oswaldo Cruz was appointed to effect the changes and on March 23, 1903, was named director general of public health. Until 1909, Cruz also directed the Instituto de Manguinhos, which is why its activities were devoted to health campaigns from 1903 to 1906. Starting in 1906, the institute performed a role analogous to that of European experimental medicine centers.[107]

Developing this program, without a doubt, meant a profound change in the dominant practices of the time: health programs were devoted to the organization of services concerned with control of collective illnesses, the hegemony of the etiologic model based on control of pivotal animals and vermin, the concentration of research in state-run institutions, and the influence of the Pasteur Institute's institutional model.[108]

The second political milestone of the period occurred between 1918 and 1924 (beyond the scope of this chapter) with the birth of the country's first social politics, when the Departamento Nacional de Saúde Pública (National Department of Public Health) was established and Carlos Chagas instituted his health reforms.[109]

Notes

We are grateful for the help of María José Rueda in the preparation of this chapter.

1. Foucault, *El nacimiento*.

2. The iatrochemical system was the first modern system to confront classic theories of illness and to try to implement findings from the primary medical trends of the age. It adopted chemical interpretations of bodily functions and illness proposed by alchemists but did away with alchemy's metaphysical views, replacing them with Cartesian and atomist mechanics and Bacon's inductive method. Iatromechanics used physics to interpret the human body and its illnesses. Vitalism explained life and illness through the abstract principle of a vital force. *More botanico* nosology took as its point of departure botanical classifications, applying them to the empirical, methodological classification of illness.

3. López Piñero, *Ciencia*.

4. Ibid.

5. Foucault, *El nacimiento*, pp. 97–105.

6. Pecker, *La médicine*, p. 51.

7. Foucault, *El nacimiento*, pp. 99.

8. Pecker, *La médicine*, p. 51.

9. López Piñero, *Ciencia*, p. 13.

10. Ibid.

11. Martínez Cortés, *La medicina*.

12. In the context of Galenic medicine, illness has two phases: the *nosos*, or potential risk of sickness; and the *pathos*, or de facto illness. Nosology studies typical forms of human illness; pathology studies concrete illness.

13. López Piñero, *Ciencia*, p. 29.

14. Ibid., pp. 30–31.

15. Systematic observation originated with Hippocratic medicine. Palpating as a medical practice had been used since the Middle Ages, but percussion and auscultation were introduced later: Corvisart, supported by Leopold Auenbrugger's *Inventum Novum*, began the practice of percussion, and Laennec, following in Bayle's footsteps, began mediate auscultation.

16. Miranda, "Apuntes," p. 151.

17. Ackerknecht, *Medicine*.

18. López Piñero, *Ciencia*, p. 13.
19. Ibid., p. 47.
20. Ibid., pp. 47–49.
21. Ibid., p. 50.
22. To these investigators, hospitals were no more than the vestibule of medicine; the laboratory was its true sanctuary.
23. The Comtian trend in positivism (begun in France around 1826, when Auguste Comte began to teach positive philosophy), is clearly linked to the developments in experimental physiopathology and goes beyond empiricism and sensationalism, though it is rooted in both. According to this philosophy, reason is used as an instrument with which to order and understand data contributed by facts, through control allowed by experimentation and controlled experience, not as a method of speculation or contemplation to reach essential truths.
24. López Piñero, *Ciencia*, pp. 95–120.
25. Ibid., pp. 48–51.
26. Beginning in the early nineteenth century, the first physiopathologists (e.g., François Magendie, Pierre Flourens, and Johannes Müller) began a polemic against hospital-based or anatomical and clinical medicine whose foundation was—as we have noted—on sensationalist empiricism.
27. This approach conceived of illness as an alteration of the body's normal functions, the process and causes of which must be studied not from the viewpoint of Condillac's Enlightenment empiricism or sensationalism but with a purely experimental methodology and point of view centered on laboratory work. See Laín Entralgo, *La historia clínica*.
28. Laín Entralgo, *El diagnóstico médico*, pp. 85–91; idem, *Pasteur*, p. 14.
29. Arboleda, "Acerca del problema de la difusión científica," pp. 3–4; Quevedo, *La institucionalización*.
30. Quevedo and Zaldúa, "Antecedentes."
31. Peset and Lafuente, "Un modelo."
32. Quevedo, *La institucionalización*.
33. Cabrera Alfonso, "La academia," p. 71; Flores, *Historia*, p. 151.
34. Gortari, *La ciencia*, p. 274.
35. Even though Quito had a fairly well educated physician—Dr. Eugenio de Santa Cruz y Espejo—who espoused Enlightenment ideals, his liberal posture prevented his active linking to reforms in medical teaching; his role was more to spread and research the modern Sydenhamiam medical paradigm (see Albarracín, "La medicina colonial").
36. Aceves and Saldaña, "La cátedra," p. 211.
37. Gortari, *La ciencia*, p. 248.
38. Flores, 1888, p. 152.
39. A group of physicians espoused Enlightenment ideals; among them, José Ignacio Bartolache, José Mariano Mociño, and Luis Montaña fought for the promotion of Enlightenment ideas. They did not, however, achieve any significant change in the university's viewpoint. Montaña was finally appointed to the Vísperas Chair in Medicine in 1815, from which he tried to teach modern clinical science.
40. The Facultad de Medicina, with the exception of Montaña's Vísperas class period, kept teaching Hippocrates, Galen, and Avicena until the late nineteenth

century. Not until 1824 were new texts by Enlightenment authors (La Cava Villaverde and La Falle) introduced in anatomy classes to replace those by Galen.

41. Flores Troncoso, *Historia*, p. 109; Martínez Cortés, *La medicina*, pp. 59–75.
42. Scientific internationalization passed through borders by ad hoc routes and generated its own informal spaces at certain junctures. The case of Mexico, as we will see later, was rather significant.
43. Texeira y Dantas, *História*, p. 76.
44. Santos Filho, *História*, pp. 290–291.
45. Santos Filho, *Pequena história*, p. 76.
46. Texeira y Dantas, *História*, pp. 176–182; Da Silva, "Tramissão," p. 145.
47. This position did not exist in Latin America.
48. López Piñero, *Ciencia*, p. 80.
49. Quevedo y Zaldúa, "Antecedentes."
50. Quevedo y Vergara, "El proceso."
51. Quevedo and Zaldúa, "Institucionalización," pp. 279–280.
52. Miranda Canal, "Apuntes."
53. Quevedo y Vergara, "El proceso," pp. 41–66.
54. *La Lanceta* was a journal of medicine, surgery, natural history, chemistry, and pharmacy that began publication in Bogotá in April 1852.
55. See *Gaceta Médica de Colombia*, which began publication in Bogotá on July 6, 1864.
56. Miranda, "Apuntes," p. 157.
57. Ibid., pp. 163, 181.
58. Cosío Villegas, *Historia*, pp. 83–92.
59. Martínez Cortés, *La medicina*, pp. 42–48.
60. Téllez and González Bonilla, "La influencia francesa."
61. Martínez Cortés, *La medicina*, pp. 42–48.
62. Ibid., pp. 68–73.
63. Santos Filho, *Pequena história*, pp. 76–82; Da Silva, "Tramissão," p. 145.
64. Da Silva, "Tramissão."
65. Schwartzman, *Formação*, pp. 66–67.
66. Da Silva, "Tramissão."
67. Schwartzman, *Formação*, pp. 67–68.
68. Ibid., p. 69.
69. Obregón, "El sentimiento."
70. Miranda, "Apuntes," p. 155.
71. Ibid., pp. 170–178.
72. Ibid., p. 165.
73. De Asúa, "Influencia," pp. 79–89.
74. Serna de Londoño, *Anotaciones*, pp. 72–84.
75. Miranda Canal, "La medicina."
76. Jaramillo Uribe, *El pensamiento*.
77. García, *El método experimental*.
78. Luz, *Medicina*, p. 108.
79. Ibid., p. 106.

80. Ibid., pp. 112–113.
81. Stepan, *Génese*, pp. 63–64.
82. Luz, *Medicina*, p. 114.
83. Ibid.
84. Stepan, *Génese*, p. 61
85. Schwartzman, *Formação*, p. 70.
86. Peard, *Tropical Medicine*.
87. Luz, *Medicina*, p. 107.
88. Stepan, *Génese*, pp. 83–100.
89. Martínez Cortés, *La medicina*, p. 74.
90. Ibid., pp. 84–88.
91. Ibid., pp. 129–137.
92. Ibid., pp. 145–149.
93. Howard-Jones, *Antecedentes*.
94. Howard-Jones, "La Organización Panamericana de la Salud."
95. Quoted by N. Howard-Jones, "Problemas."
96. Ibid.
97. Ibid.
98. Quevedo, Miranda, Mariño, Hernández, Wiesner, Cárdenas, *La salud*.
99. The bibliography on the polemical Rafael Núñez is comparatively extended. For his political and philosophical ideas, see Jaramillo Uribe, *El pensamiento colombiano*.
100. Osorio, "Informe," p. 147.
101. García Medina, *Compilación*, pp. 3–21.
102. Mesa, "La vida política," p. 88.
103. Martínez Cortés, *La medicina*, pp. 137–141.
104. Luz, *Medicina*, pp. 121–127.
105. Costa, *Lutas*, p. 34.
106. Ibid., pp. 12–17.
107. Benchimol, *Manguinhos*, pp. 23–47.
108. Costa, *Lutas*, p. 35.
109. Ibid., p. 15.

Bibliography

Aceves Pastrana, P., and J. J. Saldaña. "La cátedra de botánica y los gremios de la medicina en el Real Jardín Botánico." In *Memorias del Primer Congreso Mexicano de Historia de la Ciencia y la Tecnología*, vol. 1, J. J. Saldaña (ed.). Mexico City: Sociedad Mexicana de Historia de la Ciencia y la Tecnología, 1987.

Ackerknecht, E. *Medicine at the Paris Hospital (1794–1848)*. Baltimore, Md., 1967.

Albarracín Teulón, A. "La medicina colonial en el siglo XVIII: De los Aires, Aguas y Lugares hipocráticos a las reflexiones higiénicas del ecuatoriano Eugenio Espejo." *Asclepio* 29, no. 2 (1987): 2–26.

Arboleda, L. C. "Acerca del problema de la difusión científica en la periferia: El caso de la física newtoniana en la Nueva Granada, 1740–1820." *Ideas y Valores, Revista Colombiana de Filosofía* 79 (1989): 2–26.

———. "José Celestino Mutis et la formation d'une tradition scientifique en la Nouvelle Granade." In *Naissance et développement de la science-monde*, X. Polanco (ed.). Paris, 1990.

———. "Mutis entre las matemáticas y la historia natural." In *Historia social de las ciencias: Sabios, médicos y boticarios*, D. Obregón (ed.). Bogotá: Universidad Nacional de Colombia, 1986.

Basalla, G. "The Spread of Western Science." *Science* 156 (May 5, 1967): 611–622.

Benchimol, J. (coord.). *Manguinhos do sonho á vida: A ciência na Belle Époque*. Rio de Janeiro: Casa de Oswaldo Cruz, 1990.

Braudel, F. *Civilización material, economía y capitalismo, siglos XV–XVIII*, vol. 3. Madrid: Alianza Editorial, 1979.

Cabrera Alfonso, J. R. "La academia de anatomía práctica de México." In *Anales de las II Jornadas de Historia de la Medicina Hispanoamericana*. Cádiz: University of Cádiz, 1986.

Canguilhem, G. *Ideología e racionalidade nas ciências da vida*. Lisbon, 1977.

Cosío Villegas, D., et al. *Historia mínima de México*. Mexico City: El Colegio de México, 1974.

Costa, N. do R. *Lutas urbanas o controle sanitário*. Rio de Janeiro, 1986.

Da Silva, M. B. N. "Tramissão, conservação e difussão da cultura no Rio de Janeiro (1808–1821)." *Revista de Historia* 47, no. 97 (1974): 137–159.

De Asúa, M. J. C. "Influencia de la Facultad de Medicina de París sobre la de Buenos Aires." *Quipu, Revista Latinoamericana de Historia de las Ciencias y la Tecnología* 3, no. 1 (1986): 79–89.

Flores Troncoso, F. de A. *Historia de la medicina en México (1888)*. Facsimile. Mexico City: Instituto Mexicano del Seguro Social, 1982.

Foucault, M. *El nacimiento de la clínica: Una arqueología de la mirada médica*. Madrid, 1975.

García Medina, P. *Compilación de las leyes, decretos, acuerdos y resoluciones vigentes sobre higiene y sanidad en Colombia*. Bogota, 1923.

———. *El método experimental aplicado a la clínica médica*. Bogotá: Imprenta de la Luz, 1897.

Gortari, E. *La ciencia en la historia de México*. Mexico City: FCE, 1963.

Howard-Jones, N. *Antecedentes científicos de las conferencias sanitarias internacionales, 1851–1938*. Geneva: World Health Organization, 1975.

———. "La Organización Panamericana de la Salud: Orígenes y evolución." Pt. 1. *Crónica de la OMS* 34 (1980): 395–403.

———. "Problemas de organización de la salud pública internacional entre las dos guerras mundiales." Pt. 1. *Crónica de la OMS* 31 (1977): 452–459.

Inkster, I. "Scientific Enterprise and the Colonial 'Model': Observations on Australian Experience in Historical Context." *Social Studies of Science* 15 (1985): 677–704.

Jaramillo Uribe, J. *El pensamiento colombiano en el siglo XIX*. Bogotá: Editorial Temis, 1982.

Lafuente, A., and J. Sala Catalá. "Ciencia colonial y roles profesionales en la América española del siglo XVIII." *Quipu, Revista Latinoamericana de Historia de las Ciencias y la Tecnología* 6, no. 3 (1989): 387–403.

Laín Entralgo, P. *El diagnóstico médico: Historia y teoría.* Barcelona: Salvat, 2982.
———. *La historia clínica: Historia y teoría del relato patológico.* Barcelona: Salvat, 1961.
———. "Pasteur en la historia." In *Pasteur*, vol. 1, R. J. Dubos (ed.), pp. 9–17. Barcelona: Salvat, 1985.
López Piñero, J. M. *Ciencia y enfermedad en el siglo XIX.* Madrid, 1985.
———. *Ciencia y técnica en la sociedad española de los siglos XVI y XVII.* Barcelona: Labor, 1979.
Luz, M. T. *Medicina e ordem política brasileira.* Rio de Janeiro, 1982.
Macleod, R. "On Visiting the 'Moving Metropolis': Reflections on the Architecture of Imperial Science." In *Nuevas tendencias en historia de la ciencia*, A. Lafuente and J. J. Saldaña (eds.). Madrid: Consejo Superior de Investigaciones Científicas, 1978.
Martínez Cortés, F. *La medicina científica y el siglo XIX mexicano.* Mexico City: Siglo XXI/SEP, 1987.
Mesa, D. "La vida política en Colombia después de Panamá." In *Manual de historia de Colombia*, J. Jaramillo Uribe (ed.). Vol. 3. Bogotá: Instituto Colombiano de Cultura, 1980.
Miranda Canal, N. "Apuntes para la historia de la medicina en Colombia." *Ciencia, Tecnología y Desarrollo* 8, nos. 1–4 (1984): 121–209.
———. "La medicina colombiana de la Regeneración a los años de la Segunda Guerra Mundial." In *Nueva historia de Colombia*, A. Tirado Mejía (ed.), vol. 4, pp. 257–284. Bogotá: Planeta, 1989.
Obregón, D. "El sentimiento de nación en la literatura médica y naturalista de finales del siglo XIX en Colombia." *Anuario de Historia Social y de la Cultura*, nos. 16–17 (Bogotá) (1990): 141–161.
Osorio, N. "Informe a la junta central de higiene de Bogotá, sobre las cuarentenas." *Revista de Higiene* 1, no. 7 (September 1888): 143–150.
Peard, J. G. "Tropical Medicine, Society and Race: The Case of the Escola Tropicalista Bahiana, Brazil, 1860–1889." In *Mundialización de la ciencia y cultura nacional*, A. Lafuente, A. Elena, and M. L. Ortega (eds.), pp. 563–572. Madrid: Doce Calles, 1993.
Pecker, A. *La médicine à Paris de XIII au XX siècle.* Paris: Éditions Herras, 1984.
Peset, J. L., and A. Lafuente. "Un modelo de ciencia aplicada: El conocimiento de la naturaleza en la España ilustrada." Unpublished. N.d.
Polanco, X. "La ciencia como ficción: Historia y contexto." In *El perfil de la ciencia en América*, J. J. Saldaña (ed.). *Cuadernos de Quipu* 1 (1987): 41–56.
———. "One science-monde: La modialisation de la science européenne et la création de traditions scientifiques locales." In *Naissance et développement de la science-monde*, X. Polanco (ed.). Paris: La découvert/Conseil de l'Europe/UNESCO, 1990.
Pyenson, L. "Ciencia pura y hegemonía política: Investigadores franceses y alemanes en Latinoamérica." In *Nuevas tendencias en historia de la ciencia*, A. Lafuente, and J. J. Saldaña (eds.). Madrid: CSIC, 1987.
———. "*In Partibus Infidelium*: Imperialist Rivalries and Exact Sciences in Early Twentieth-century Argentina." *Quipu, Revista Latinoamericana de Historia de las Ciencias y la Tecnología* 1, no. 2 (1984): 253–303.

Quevedo, E. "José Celestino Mutis y la educación médica en el Nuevo Reino de Granada." *Ciencia, Tecnología y Desarrollo* 8, nos. 1–4 (1984): 405–415.

———. "La institucionalización de la educación médica en la América hispano-lusitana." *Quipu, Revista Latinoamericana de Historia de las Ciencias y la Tecnología* 10, no. 2 (1993): 165–188.

Quevedo, E.; N. Miranda; C. Mariño; M. Hernández; C. Wiesner; and H. Cárdenas. *La salud en Colombia: Análisis socio-histórico*. Bogotá, 1990.

Quevedo, E., and A. Vergara. "El proceso de institucionalización de la educación médica en Colombia." *Revista Escuela Colombiana de Medicina* 1, no. 1 (1988): 41–66.

Quevedo, E., and A. Zaldúa. "Antecedentes de las reformas médicas del siglo XVIII y el XIX en el Nuevo Reino de Granada: Una polémica entre médicos y cirujanos." *Quipu, Revista Latinoamericana de Historia de las Ciencias y la Tecnología* 3, no. 3 (1986).

———. "Institucionalización de la medicina en Colombia." Pt. 2. *Ciencia, Tecnología y Desarrollo* 13, nos. 1–4 (1989): 233–310.

Saldaña, J. J. "La formation des communautés scientifiques au Mexique (du XVI au XX siècle)." In *Naissance et développement de la science-monde*, X. Polanco (ed.). Paris: La découvert/Conseil de l'Europe/UNESCO, 1990.

———. "Nacionalismo y ciencia ilustrada en América." In *Ciencia, técnica y estado en la España ilustrada*, J. Fernández Pérez and I. González Tascón (eds.). Zaragoza, 1990.

Santos Filho, L. *História geral de medicina brasileira*, vol. 1. São Paulo, 1977.

———. *Pequena história da medicina brasileira*. São Paulo, 1980.

Schwartzman, S. *Formação da comunidade científica no Brasil*. Rio de Janeiro: Editora Nacional, 1979.

Serna de Londoño, C. *Anotaciones sobre la historia de la medicina en Antioquia*. Medellín, 1984.

Stepan, N. L. "Eugenesia, genética y salud pública: El movimiento eugenésico brasileño y mundial." *Quipu, Revista Latinoamericana de Historia de las Ciencias y la Tecnología* 2, no. 3 (1985): 351–384.

———. *Génese e evolução da ciência brasileira*. Rio de Janeiro, 1976.

Téllez, E., and G. González Bonilla. "La influencia francesa en el desarrollo de las ciencias médicas en México." In *Memorias del Primer Congreso Mexicano de Historia de la Ciencia y la Tecnología*, vol. 1. Mexico City, 1989.

Texeira, F. M. P., and J. Dantas. *História do Brasil, da colônia á república*. São Paulo, 1979.

CHAPTER 7

Academic Science in Twentieth-century Latin America

HEBE M. C. VESSURI

Introduction

This chapter analyzes the period during which a scientific community began to emerge in Latin America; it also explores in some detail the advocacy and organization of science in the region since the end of the nineteenth century. This journey gives us the opportunity to come into contact with a wide spectrum of issues that were previously scattered in a variety of articles and monographs. These issues include the importance of scientific societies; private and official sponsorship; conflicting bureaucratic and intellectual notions regarding research, particularly when the state took charge of organizing scientific activity; and the slow process of decision making, which, since the nineteenth century, has been related to economic survival, industrial development, and scientific progress.

The emergence of the scientific community can be divided into five periods. During the first phase, modern science appeared in the region with close links to the principles of European positivism, a key element of the political and economic modernization schemes of the new nations at the end of the nineteenth century and beginning of the twentieth. The second phase is marked by the newborn institutionalization of experimental science (1918–1940). The third phase can be described as the development decades (1940–1960). The fourth phase might be called the Age of Scientific Policy (1960–1980). Finally, the fifth phase witnessed the appearance of a new audience for science: industry (1980–1990).

Behind this account lies what I call the incorporation/autonomy thesis. This is a simple theoretical scheme that aims to provide a coherent explanation of selected and interpreted data that would otherwise be arranged arbitrarily. My objective is to develop a sociological explanation of a thesis and not, given the scope and the available space, to provide a historical analysis. Several reasons have led me to the incorporation/autonomy thesis as a starting point. First, I consider it a fruitful methodological approach to the conceptualization and interpretation problems inherent in the available historical data. Second, when a sociology-of-science approach is adopted, some of the basic contradictions of historical and sociological theory can be resolved. Third, the thesis is appealing for analyzing both social change and problems regarding the relative weight of intellectual and practical determinants in the process of integrating scientific concepts, instruments, themes, institutions, and patterns. Finally, a comparative approach to the social study of science introduces the dimensions of political and economic power, diplomacy, and cultural heterogeneity.

The explanation must remain highly simplified, although it encompasses several analytical levels. In cognitive terms, it can be argued that accounting for the dominance of one notion of science or another at a given moment in time means considering an important component of Latin American thinking: the tendencies and polarized controversy surrounding founders, followers, and precursors. It involves schools, families, dissidence, and opposition. Its history necessarily includes not only the production of natural and social scientists but also that of some philosophers, writers, and artists from Latin America, Europe, the United States, and other countries.[1]

The trends in Latin American and national thinking have always intermingled with contemporary theories, themes, and intellectual trends from Europe and the United States. But then again, the basic concerns of Latin American and foreign science have not always been contemporary, nor have they flourished and died at the same time. Rhythms have often been out of sync. But there has always existed a wide exchange of concepts and themes, without which it is impossible to understand how Latin American science organized, developed, destroyed, and re-created itself.

It is obvious that Latin American science has not grown in a continuous, harmonious way. On the contrary, there have been advances and setbacks. Innovative contributions were questioned and rejected; others reappeared from the past and were presented as new, just dressed in a different language. The dynamics of controversial theories and social forces induced frequent reorientation.

Within the institutional context, it is important to point out the framework within which the transfer and reproduction of knowledge took place, how institutions were affected by cognitive change, and how, at the same time, different cognitive criteria were mediated by particular institutions.[2]

I am assuming that the development of institutional context does not simultaneously follow cognitive change but reflects an already changed pattern. Scientific activity was organized at different moments by prevailing institutional contexts: the university; the institute dedicated exclusively to research; the science museum; the observatory; the scientific journal; and so on. The internal configuration of these ambits and their mutual relationships were determined by the university's centrality during most of the period with which we are concerned. It was not until the 1960s that the university's central position began to erode and the conditions were established for the science and technology sector to be reconstructed in the 1970s. The various institutional models did not grow spontaneously; they were the result of ideological positions that were defined by the clash of intellectual struggle, the commitment to a social context, and the adoption of plans that were drawn up in other parts of the world. The weight each of these factors had on the institutional response and how they influenced the history of research and of the institutions themselves are matters that can be made clear through specialized research.

I will also point out the state's changing function. It has played, in different ways and with varying efficiency, a leading role in matters related to scientific activity in Latin America, although it also performed some obscurantist interventions. The state is the link that has tied together most of the scientific activity in Latin America. It has accomplished this by monopolizing higher education in public universities—a fact that remained unquestioned for a long time by social representatives and intellectuals—and by creating institutes intended to carry out very specific research activities related to the productive or services sector.

The methodological basis of my argument is not entirely satisfactory, mainly because selection of particular cases that support it can be problematic. My periodization is one of the problems. A second problem, related to my periodization, is that each period is illustrated by only some cases of national development. It is clear, though, that all countries had processes that were sometimes similar; this similarity helps describe the characteristics that distinguish a country's development pattern from that of all others in the region. Obviously, my intention is not to revise thoroughly any particular country's history of science; thus, many important aspects will remain untouched. I will therefore be satisfied if the reader does not discard my obviously preliminary contribution, with its new hypotheses and adjustments to old ones, as demonstrably false or irrelevant.

Order and Progress: Positivist Science and the Arrival of the Twentieth Century

Toward the end of the nineteenth century, Latin America's scientists belonged to a small group of scholars and cultivated persons among whom experimental researchers, naturalists, and amateurs were not yet clearly defined.[3] Nonetheless, institutional and cognitive changes were taking place as a consequence of the social transformation that was shocking the new Latin American nations and that modified established scientific enterprise. Most countries had begun a process of economic and political reconstruction that was marked by the expansion of the export economy and the consolidation of civil oligarchies. European positivism was enthusiastically welcomed, partly for political reasons, for it offered politicians and intellectuals a conceptual scheme that allowed society and history to be placed within a framework of progress. It also reinforced an "official ideology" of the civil elite that stressed the uselessness of revolts, military governments, and the church for assuring the order and stability necessary for society to evolve naturally. In addition, it pointed out the need for development that was based on free enterprise.

Science, education, European immigration, and foreign investment were considered the main instruments for rebuilding Latin America according to modern patterns.[4]

Within the realm of thought, progress was considered to consist of acquisition of knowledge that was produced mainly in other parts of the world. The "new" model of positivist science was basically a new method for organizing and systematizing experience and, thence, knowledge. European knowledge and technology were perceived as necessary for national development. Scientific studies, expeditions to the interior, and inventories of the native flora and fauna were seen as effective ways to gain true understanding of a new nation's resources and possibilities.

But such interests were not merely a product of educational reform; other factors played a role as well, such as the interest foreign enterprises showed in controlling Latin American resources. Brockway (*Science*) analyzes the role some scientific institutions played in the spread of imperialism in modern world history. The historical differences between advanced nations such as the United Kingdom and the new, fragile countries, such as those in Latin America, endowed the former with great advantages over the latter in terms of power, prosperity, and stability. This being so, it was only "natural" for Great Britain to think it could break the law and unity of countries such as Brazil, Mexico, or the Andean republics by instructing its consuls to steal seeds that were to be deposited and investigated at the Royal Botanic Gardens at Kew and then redistributed among the empire's eastern colonies.

Inspired by positivist science, Latin America encouraged the collection of a great amount of data.[5] Some areas saw the development of texts: explanations of infinitesimal calculus by Francisco Díaz Covarrubias, Manuel Gargollo, and Manuel Ramírez from Mexico; the geological and paleontological works of Florentino Ameghino and Karl Hermann Burmeister; Benjamin Apthorp Gould's astronomical contributions in Argentina; Hermann von Ihering's and Oswaldo Goeldi's natural science texts in Brazil.[6] These texts made European knowledge available to Latin American students—even if only in a few disciplines. The lack of a true historiography of contemporary science in Latin America makes it difficult to assess in any but the philosophical aspect, not to mention compare, how positivism was received in each country.[7] In general, the conventional wisdom is that positivism provided a social appreciation for science as a source of progress and practical knowledge, but limited to mere rhetoric in support of research and only occasionally becoming persistent research.

A promising research area is the study of the controversies and conflicts between European scientists and local scholars regarding the legitimacy of certain themes and theories in areas where local scientific capacity had been created. Often, these conflicts originated from the arrogance and rigidity of the foreign professors. Some of them, as Safford notes with regard to Colombia,[8] came to Latin America with a superior European attitude and tended to consider the university laboratories (financed by the state) as their property and used them for all kinds of personal projects. They demanded that their laboratories be maintained and supplied like those in Europe, and when the university rectors could not keep up with their demands, the imported professors were frequently very disrespectful. The struggle for legitimacy and power between Argentine evolutionary paleontologist Ameghino and German creationist zoologist Karl Burmeister is very illustrative.[9]

Argentine paleontology had reached critical mass, contributing an original approach to evolution studies. Among its signs of maturity by the end of the nineteenth century we can point to an interdisciplinary group, Darwinist control of one of the two most important museums in Argentina, support from the Ministerio de Educación (Ministry of Education), and multiple contacts with the European research community. Ameghino's early works were published in France and the United States, and he kept up close contact—including active collaboration with Henri

Gervais—with the icons of French transformism.[10] Ameghino gained nationalist momentum for his cause, which helped him get support and counter the traditionalist opposition.[11]

Another illustrative controversy concerns the 1913 debate between Peruvian bacteriologists and a mission of North American physicians headed by Richard Strong, from Harvard University, to study the causes of Oroya fever. The Peruvian scientists considered Strong's conclusions not only mistaken but offensive to national medical knowledge. Driven by a nationalist spirit, they searched until 1925 for the empirical evidence they needed to prove their single-cause theory; a Japanese researcher from the Rockefeller Institute completed the experiment.[12]

Altitude physiology also reflected Peruvian indigenism, for Andean biology was part of a national discussion about the role of natives in Peruvian society. By the 1920s, the debate between Monge Medrano and British physiologist Joseph Barcroft, from Cambridge University, on the "normality" of the Andean was clear proof of the originality of Latin American scientific thinking.[13]

All of these examples show how hard it was, even locally, to create adequate institutional space where scientists and their theories could grow strong. In addition, legitimization always depended on representatives from scientific centers in the advanced countries.

The restructured university, the observatory, and the natural sciences museum were the central institutions of the new "positivist science," and their functions and internal structure reflect the concept of science at that time.[14] A glimpse at the institutional picture reveals how the different socially pertinent criteria still prevailed and how science had to replace the existing and firmly institutionalized systems of knowledge (especially the religious ones) to become the dominant way of learning and experimenting. The number of persons who could take charge of research in Latin America was small, and something close to a research chair was almost totally absent.

The contrast with Europe was shocking. At the universities, science was subordinated to liberal education. Science was taught because it kept the mind disciplined, but most of the scientific subjects did not go beyond the elementary level. Students had to learn physics and mathematics not to become scientists or engineers but to get a good basic education. With few exceptions, science was taught by reading and recitation. Students very rarely entered the laboratory to question nature; instead, they learned from textbooks what science had to say. It was common for teachers to be priests who answered to the church hierarchy.

Although Argentine and Brazilian engineers dominated the different branches of civil engineering starting in the last quarter of the nineteenth century,[15] it was not uncommon to find that the emphasis on practical aspects in the engineering schools was somehow inhibited by the general ideological climate. To Safford,[16] the lack of a spirit of enterprise in the late-nineteenth-century engineer can be explained by a tendency to think of himself as a professional who defined his functions according to prevailing social values. Engineers in Bogotá seem to have considered themselves as public functionaries, scholars, and technical experts, but most of them could not perceive of themselves as businessmen, not even potentially. The *Anales de Ingeniería* (Engineering Annals) began addressing some members of the fraternity of engineers as "doctor" when engineers in Bogotá started to incorporate themselves into the capital's prevailing political pattern.[17]

Latin American students with scientific or medical ambitions felt increasingly lured to study abroad, particularly in France, Germany, or the United States. Oswaldo Cruz is a good representative of the new intellectual scene, adaptable to the European institutional model (the Pasteur Institute) and to Brazilian conditions (the Instituto Manguinhos in Rio de Janeiro).[18]

A renewed interest in science and technology, as well as an admiration for European culture, allowed observation and experimentation to become legitimate activities in some universities. Also, the new intellectual climate led to a concern for undergraduate education. In 1904, the Instituto Nacional del Profesorado Secundario (National Institute of High School Teachers) of Buenos Aires was founded according to Prussian educational models, in order to remedy the lack of specialized teachers. Between 1904 and 1913, some twenty foreign teachers were hired—most of them German.[19]

The process of institutionalizing science laboratories—which was still going on in the international context—gained new force with governmental support, since governments were interested in public health and agricultural production.[20] Vocational education was restricted to a few technological institutions. In Argentina, the earliest industrial studies began in 1898, when an industrial department headed by engineer Otto Krause became part of the Escuela de Comercio (Business School) of Buenos Aires. Later, the department became the Escuela Nacional Industrial (National Industrial School), renowned for its first director, Otto Krauze, and its crucial importance to the development of technical education.[21]

By the early twentieth century, many graduates in medicine, law, and engineering held positions in societies that were growing in complexity. Individuals with scientific education were rapidly absorbed by the state bureaucracy.[22] A new scientific market was born.[23]

Nonetheless, what happened when science became a public issue was not a simple process. In many ways, it meant the expansion of a profession that was wide open to only moderately talented individuals and growing dependence on an administration little concerned about intellect and reluctant to give scientists the independence they needed—all in the holy name of democracy. But this is just part of the story, for the state was the sponsor of all research activity and even of educational programs in universities and other institutions of learning.

Educational structures at all levels underwent some profound changes. One of the factors that marked the difference between Latin American countries during the last quarter of the nineteenth century was the exclusive or inclusive nature of access to education as a way of ideological imposition. Argentina, along with Uruguay, Costa Rica, and, to a lesser degree, Chile, joined the international market as exporters of raw materials and as importers of manufactured goods; all were socially and legally organized to assume that everyone was included in the basic circles of cultural diffusion. Most of the population had access only to the most basic education, which guaranteed cultural homogeneity; a selected elite had access to more expanded intellectual expressions and to the dominion of educational instruments, which allowed for a certain degree of knowledge creation. Democracy became institutionalized via the public schools, which, under the slogan "Universal Education," was a guarantee of homogeneity in the educational process. Education's purpose was to teach secular values, republican principles, and a certain scientific vision of reality, which expressed—with a relatively high degree of correspondence—the cultural order that marked the most dynamic social sectors. Its fundamental task was to give form to a model citizen (both a leader and a follower) within the parameters of liberal democracy.

In contrast, other countries deepened the exclusionary and restrictive access to education, which, in the most extreme cases, had negative consequences that still endure. Brazil, for instance, still has a high illiteracy and elementary school drop-out rate and has not been able to solve the problem of marginalization in large sectors of the population who do not have access to the benefits of culture diffusion.[24] Equal access to education continues to be a problem in much of Latin America. The Southern Cone's advantage in having an educated population in the first half of the century was undoubtedly a necessary condition for economic and industrial growth, although it was not enough to maintain and reinforce the socioeconomic and political development pattern. Those countries intensified their early modernization process by following certain patterns in order to incorporate themselves into the international economic, cultural, and scientific system.

The Foundations of Experimental Science, 1918–1940

With World War I, the enthusiasm for positivism began to fade. The ideal of progress had not come true, and order had been understood as maintaining the status quo. Also, positivism's romantic science and development promise weakened when Latin American intellectuals found progressive Europe involved in a bloody war: World War I. The Old World's leaders were incapable of keeping the ideals of peace and progress alive, which deepened disappointment in positivism. In addition, intellectuals and politicians discovered through experience the difference between scientific research and industrial applications of science and technology. The former did not always have material results, and the latter could well be developed through empirical knowledge alone.

The period between wars witnessed the profound transformation of Latin American society. Union strikes and student revolts provided the background. A new stage in working-class political organization was marked by the appearance of the Communist and Socialist parties in countries like Argentina and Brazil. The revival of Catholic thinking, especially the defense of religious education, pervaded the region. The armies of several countries were professionalized.

Intellectual production grew significantly during the period. The ascension of a new middle-class society, which was concerned with national problems, brought a whole new market for the Latin American authors and encouraged the expansion of the publishing industry. This growth in the publishing field was crucial for the professionalization and autonomy of intellectual work in Argentina, Brazil, and Mexico.

Around 1918, Buenos Aires, the gateway to the "world's granary," was the second-largest Atlantic city, after New York. No other country in the world imported as much merchandise per capita as did Argentina, with the exception of commercial import and distribution centers like the Netherlands and Belgium. In 1911, Argentina's per capita income was greater than Canada's and a fourth that of the United States. In 1914, the per capita income in Argentina was equal to that of Germany and the Netherlands and higher than that of Italy, Spain, Switzerland, and Sweden and had grown at an annual rate of 6.5 percent since 1869.

The traditional Argentine universities were ripe to produce their own internal transformation. There were signs of the ongoing modernization process at institutions such as the Universidad Nacional de La Plata, where, in the first half of the twentieth century, the efforts of a number of German physicists and astronomers made this institution one of the best Latin American schools for the exact sciences. The challenges posed by the dramatic transformation of the social and economic realm could not be met by most of the existing universities. They were imprisoned by the "degenerate Creole" version of the Napoleonic professional pattern,[25] and their teaching methods were colonial dead weight.

A growing frustration began to take shape within the professorial ranks and among students, and a program of university reform was begun in 1918. All of the university community in Argentina and the rest of Latin America adhered to it.[26]

In Mexico, the Universidad Nacional had been in existence since the 1910 Revolution but on a basis that was different from that of the Real y Pontificia Universidad (Royal and Pontifical University). In fact, the Escuela de Altos Estudios (School of Higher Education)—also established in 1910—was the predecessor of the Facultad de Ciencias (Science Department), created in 1939. With the creation of several other institutes, scientific research would gain strength.

Agricultural research was organized at the Secretaría de Agricultura (Ministry of Agriculture) in the 1930s. The revision of the medical curriculum began in 1922 at the Hospital General when gastroenterology was reoriented toward a clinical physiology approach. The cardiology service, established in 1924 by Ignacio Chávez at the Hospital General, gave rise to the Instituto Nacional de Cardiología (National Institute of Cardiology) and to the Sociedad Interamericana de Cardiología (Inter-American Cardiology Society) in 1944. The movement would result, in the late 1940s, in Arturo Rosenblueth's research on the physiology of the nervous and cardiovascular system, which began at Harvard University during Rosenblueth's long collaboration with W. B. Cannon.

In 1915, the Escuela de Artes y Oficios (School of the Arts and Trades) was transformed into the Escuela Práctica de Ingenieros, Mecánicos y Electricistas (Vocational School for Engineers, Mechanics, and Electricians). In 1916, the Escuela Química Industrial (School of Industrial Chemistry) and the Escuela Constitucionalista Médico-Militar (Constitutional

Medico-Military School) were created. In 1922, the Escuela de Salubridad e Higiene (Health and Hygiene School) was established, followed, in 1934, by the Escuela de Bacteriología (Bacteriology School) at the Universidad Gabino Barreda, which was transformed in 1936 into the Universidad Obrera (Workers' University) of Mexico and then transferred to the Instituto Politécnico Nacional (National Polytechnic Institute) as the Escuela de Ciencias Biológicas (School of Biological Sciences). In 1936, the Escuela Superior de Ingeniería Química (School of Chemical Engineering) was founded.

In Peru, the Universidad de San Marcos was modernized and the Facultad de Medicina, in particular, grew. But the fragile balance between teaching resources and the number of students finally tipped when, in 1938, a massive number of students entered the first year, and the university's laboratories could not expand to meet the demand. Teaching by memorization and the magisterial classes survived because they could adapt to the conditions that the growing number of students imposed.[27]

Scientific research became increasingly restricted to small centers of excellence. Along with the university, science societies played a role in the development and spread of science, as did the Sociedad Geológica (Geological Society) in Lima. The professionalization phenomenon is seen in Peru as well as in the rest of Latin America during these decades. Cueto (*Excelencia*) notes that the medical, law, and technological professions related to industry increased at a rate greater than the population increase, according to the 1931 census in Lima.

The present level of development in areas such as geology and geophysics is the result of efforts that go back some years. Even though these areas were especially vital to mining,[28] a crucial event took place in 1922, when the Carnegie Institution of Washington installed the Magnetic Observatory in Huancayo, thus beginning the systematic recording of geophysical data. Although the observatory's initial concern was to learn about the Earth's magnetic field, its interests widened to include other geophysical parameters, it gained worldwide prestige because of the quality of the data it provided and because of the geographic location of its facilities.[29] In spite of its transfer to the Peruvian government in 1947, the observatory has mainly depended not on government support but on international subsidies to develop its research programs.

In Colombia, the first decades of the twentieth century meant relatively rapid growth. In an atmosphere of unprecedented prosperity, the industrial and urban development that took place after 1904 provided great opportunities for Colombian professionals, who created associations, established journals, and tried to gain legitimacy by supporting modern science. The nationalization of the Colombian railroads was backed by the Sociedad Colombiana de Ingenieros (Colombian Engineers' Society) after 1911.

The engineering fraternity had a double interest in the nationalization. Government control would mean more jobs for Creole engineers than private foreign control did. As engineers they were also interested in government control as a way of streamlining the railroad system. In 1915, engineer Felipe S. Escobar—supported by the engineering society of Bogotá—presented to the Senate the first comprehensive project to unify the railroads under national control.

As the economy grew, local engineers faced the problem of specialization. Before 1935, Colombia's technical schools could offer only a single, basic degree in mathematics and civil engineering. Colombian engineers wanting to specialize had to go abroad.[30]

Uruguay entered the twentieth century with a single institutional agent that actively participated in national scientific life: the university. The Museo de Historia Natural had suffered, since its creation in 1892, chronic difficulties that kept it from playing a major role in scientific and technological development—in spite of its having had renowned directors such as Carlos Berg. The introduction of the notion of academic research was mainly the contribution of Clemente Estable, who made a deep impression on the development of biology in Uruguay. His followers founded the Instituto de Investigaciones Biológicas Clemente Estable (Clemente Estable Biological Research Institute) in 1927 to conduct basic research.[31]

Under J. V. Gómez's authoritarian regime, Venezuela experienced, from 1908 to 1927, a sui generis process of deep political repression, foreign monopolist investment in national oil, and the centralization and unification that gave birth to the modern state. Scientific and technological activity was directly related to practical concerns that led to the creation of specialized institutions: the Oficina de Sanidad Nacional (National Health Office); the Laboratorios de Bacteriología y Parasitología (Bacteriology and Parasitology Laboratory), Química Bromatológica (Chemical Bromatology), and Análisis de Agua e Investigaciones Sanitarias (Water Analysis and Health Research); the Sala Técnica (Technical Board); the Comisiones Exploradoras (Exploratory Committees) of the Ministerio de Obras Públicas (Ministry of Public Works); and the experimental agriculture stations—all of which other countries in Latin America already had in place.

The universities had problems with the regime and were open only sporadically. Many academies were created during this period, but they had a purely honorary function and were totally marginalized. On the whole, the developments were very modest.

In several areas, Chile had some interesting developments, such as within the institutional university context. Particularly between 1920 and 1940, works on biomedical disciplines such as physiology, genetics, parasitology, histology, and pathological anatomy were published, supported by figures such as Eduardo Cruz Coke and Alejandro Lipschutz. The latter supported the Sociedad de Biología (Biology Society) of Concepción, and in 1927, he established the *Boletín de la Sociedad de Biología de Concepción*. In 1928, the Sociedad de Biología of Santiago, affiliated with the society in Paris and similar to the society in Buenos Aires, was founded. Abstracts of the papers presented at its meetings are published in the *Comptes-Rendus de la Société de Biologie*. Today, biology is still the most mature and productive area.

International Cooperation

During this period between the world wars, the basis of the region's experimental science was determined and the influence of foreign professionals and institutional cooperation with advanced countries intensified. International scientific relationships were defined by Western Europe and by the United States and marked by rivalry. Both areas created similar organizations during this period.

By the late nineteenth century, the Alliance Française had been founded in France to spread culture, and the ministries had acquired administrative tools for international "intellectual" relations. After the war, colonial science institutes were created (Instituts Pasteur d'Outre-mer [Overseas Pasteur Institutes], Instituts d'Agriculture Coloniale [Colonial Agriculture Institutes], etc.).

Coming from the same general trend, the Groupement des Universités et Grandes Écoles de France pour les Relations avec l'Amérique Latine (French University and High School Group for Latin American Relations) was created in 1907.[32] From the beginning, it had two objectives: cultural influence, and competition with other countries. Although it was not totally insensitive to Latin America's "scientific needs," these were certainly not its prime objective. The Groupement's contacts were closer to diplomacy than to higher education, and with regard to the latter, it had more contacts with people in the humanities and social sciences than those in the exact sciences. Although it did not completely abandon scientific exchanges, the Groupement gradually moved from being a "scientific-cultural" project to a "diplomatic-cultural" project.

During the 1920s, it had two major activities: the creation of institutions in Latin America, and the creation of a journal, *Revue d'Amérique Latine*.[33] The Universidade de São Paulo, established in 1934, was probably the institution most influenced by the French from its founding through its early years.

The United States began working in the late nineteenth century to consolidate an empire that extended from Puerto Rico through much of Central America and on to the Philippines. The State Department's "pan-American policies" were backed by major firms, foundations, and educational institutions. Between 1913 and 1940, for instance, the Rockefeller Foundation's activities in Latin America were focused on public health and epidemic control, and Brazil got most of the funds. During this period, the foundation also developed an interest in supporting physiological research in Argentina and Peru because of the increase in high-quality work in science in those countries. The director of the Instituto de Fisiología (Physiology Institute) of the Universidad de Buenos Aires, Bernardo Houssay, in fact, was awarded the 1947 Nobel Prize for his research on the glandular basis of the metabolism of sugar.

By 1940, and coinciding with the interruption of scientific relations between the United States and Europe during World War II, the foundation stressed scientific research and support of individual Latin American researchers. The impact was significant on the region's biomedical research, and the influence of the United States grew considerably. The most important character in physiology in the United States, Walter B. Cannon, shaped many of the leaders in Latin American physiology, who went to his laboratory at Harvard on grants from the foundation. People like Efrén del Pozo and José Joaquín Izquierdo from Mexico, Juan T. Lewis and Oscar Orías from Argentina, Franklin Augusto de Moura Campos from Brazil, Joaquín V. Luco and Fernando Huidobro from Chile, and Humberto Aste-Salazar from Peru went to Harvard to work with Cannon.[34]

But the foundation was also active in other fields. Documents from 1943–1946 record the important role the foundation's Mexico field office played at the beginning of the Green Revolution, and they have recently been opened to researchers.

The Ford Foundation also made strong investments in support of the social sciences in several countries of the region. In Argentina, it helped the Instituto de Sociología de Gino Germani at the Universidad de Buenos Aires and the Instituto Di Tella during the 1950s and the 1960s. While Argentine universities were under Perón's control in 1973, active investments amounted to around $2 million in all areas and academic disciplines. During the early 1960s, the foundation's support of the social sciences in Brazil averaged $386,000 a year. It more or less doubled after the military takeover, and between 1970 and 1974, it totaled an estimated $1.2 million a year.

In Chile, where the foundation's most important program was located, close to $6 million was invested when the military ousted Allende. The Universidad de Chile alone had a budget of $10 million during the 1965–1975 period.[35]

Germany, too, made substantial incursions into Latin America, particularly in Argentina at the beginning of the century. In 1909, German officers organized the Argentine army high command, the Argentine navy began depending on German wireless technology, and German-headed firms controlled the metropolitan region's power companies. Between 1904 and 1913, the Prussian Kultusministerium (that is, the Prussian Ministry of Spiritual Affairs, Instruction, and Public Health) planned a national normal school in Buenos Aires—the Instituto Nacional del Profesorado Secundario (National Secondary School Teachers Institute)—and staffed it.[36] With the active support of the imperial Ministry of Foreign Affairs, German science and culture were implanted in open competition with the interests of the United States. The development of physics was entrusted to Emil Bose, one of the first students at the Institut für Physikalische Chemie (Göttingen Physicochemical Institute), founded by Walther Nernst.[37] In La Plata, Bose obtained the services of Johann Laub to teach electrical engineering; Laub was the first scientific collaborator of Einstein and Konrad Simons, a student of Emil Warburg's.

Bose's premature death in 1911 did not end the project in La Plata. Between 1913 and 1926, Richard Gans—who before coming to America was assistant to Nobel Prize winner Ferdinand Braun and had had a brilliant career in Tübingen and Strasbourg—supervised and approved the first six theses on physics at an Argentine university and convinced his students to publish their works in German journals. Gans's most distinguished student, Enrique Gaviola, received a PhD in 1926, in Berlin.

Astronomy followed the German pattern. Johannes Hartmann, from Göttingen, took charge of the observatory in La Plata from 1920 to 1934—his retirement year. And Einstein visited La Plata in 1925, which proves that German efforts were strong.

During this period, Spain reinforced its relationships with Hispanic America. The Institución Cultural Española (Spanish Cultural Institution), created in 1914, was the result of an effort made in 1912 by the Spanish colony in Argentina to honor the memory of Spanish researcher Marcelino Menéndez y Pelayo. Its objectives were to make public and to spread throughout Argentina scientific and literary works created in Spain through a chair that was to be held by Spanish intellectuals at the Universidad de Buenos Aires and to develop other activities that were directly related to the intellectual exchange between Spain and Argentina.[38] The institution was placed under the auspices of Madrid's Junta para Ampliación de Estudios e Investigaciones Científicas (Board for the Expansion of Scientific Research and Studies), presided over by Santiago Ramón y Cajal, "the most culturally valuable institution in Spain today, which in itself guarantees that the chair will be filled with honor, dignity, and, above all, good will."[39]

The case of Julio Rey Pastor illustrates what frequently happened with intellectuals who benefited from the Institución Cultural Española's invitations. The Spanish Civil War brought to Latin America a considerable number of Spanish intellectuals, some of whom settled temporarily and others of whom stayed for good; all of them made crucial contributions to the institutionalization of local science.[40]

Italy provided significant numbers of professors to institutions of higher education and research laboratories, at times from the great number of immigrants who flowed to Latin America for a century, sometimes as a part of official programs, whether exchange programs or specific missions, such as Luigi Buscaloni's 1899 ethnobotanical expedition to the Amazon. Students, the children of immigrants who, although born in Latin America and growing up in a mainly Italian environment, and missionaries contributed to research. A complete study of the Italian contribution to the formation of the Latin American scientific community is needed.

England played a more specific and significant role, mainly through the British Council and its grants program for Latin American students who studied specific disciplines at the renowned British universities. Luis Leloir and Maurício Oscar Rocha e Silva, for instance, studied Latin American biochemistry and pharmacology, respectively.

The Development Decades, 1940–1960

Even before World War II, theories about economic progress abounded in the region. During the 1930s and the 1940s, some scientific leaders, usually backed by international support, demanded government backing for basic research as a means for building scientific communities and for attaining economic development, both of which were thought to be causally related. As pointed out earlier, the period immediately before the war saw, in some countries, an increase in institutional activity and public policy aimed at establishing national research capacity. But the Great Depression, followed by World War II, saw a period of growth in industrial activity in Latin America that largely displaced agricultural activity, a period of rapid population expansion in the main urban centers, and a period of educational improvement, all within a political context that alternated between populism and authoritarianism.

This period had a considerable influence on the later development of the region's research activities. The notion that science and universities would play a major role in socioeconomic development was part of the *"desarrollista"* ideology from the United Nations Economic Commission for Latin America and the Caribbean (ECLAC; Comisión Económica para América Latina y el Caribe, CEPAL). Raúl Prebisch and his collaborators stressed the need to adapt to and combine international technology in order to handle Latin America's problems and to define priorities according to economic planning and to organize research to meet these priorities.[41] In practice, local manufacture was prioritized to replace imports without any concern for subsequent technological dependence. Most Latin American technology was embodied in equipment and procedures. Selecting, negotiating, acquiring, and assimilating privatized technology was totally discarded; the same happened with national research and development (R&D). Tax policies, credit practices, and the lack of payment control for technical assistance and patents rendered the cost of technology imports insignificant to the individual business owner. On the other hand, the absence of protection for the production of capital goods and the lack of support for technology investment increased the time and risk necessary for investments in technology consolidation. Objectively, investment in local technology became more expensive, which accounts for the delayed development of the capital goods sector,[42] the late start of postgraduate education,[43] the marginal structure of experimental R&D,[44] and business's low level of participation in financing these activities.[45] All of these factors contributed to shaping the region's industrial situation.[46]

Even though the general pattern of industrialization did not encourage the development of dynamic R&D systems, the strength of the modernization ideal helped university and government research gain momentum in some areas, starting in the 1950s especially.[47]

The universities were key elements of the adopted model for national scientific policy; in fact, they were the only institutions to which such a model seemed to apply in an explicit way.[48] Its purpose was to create a "scientific-technological" infrastructure that would attain some degree of critical mass, which would then automatically reinforce local technology, especially in order to exploit the development possibilities of raw materials and other national resources. All of this would increase production and productivity.

The stage for public science and technology policy, which grew during the 1960s, was set during the 1950s, and its most conspicuous spokesmen were leaders of the academic scientific community. During this period, the growth of the Brazilian scientific community was remarkable, for it began—although slowly—to expand visibly and to parallel the country's modernization and industrialization. The first Brazilian universities were not created until the 1930s.[49] A lack of qualified professors led to foreign professors being hired; it also meant a rupture with the prevailing system of higher education in the country. The Universidade de São Paulo in particular hired French professors for human sciences and German and Italian professors for physics, mathematics, chemistry, and natural sciences.[50] These foreigners shaped students and created research traditions. The Universidade de São Paulo achieved a scientific density like no other institution in the country. In 1948, the Sociedade Brasileira para o Progresso da Ciência (Brazilian Society for Scientific Progress) was created to unify Brazilian scientists and to defend and propagate scientific research;[51] its formation reflected the expansion of science in the country. In 1951, the Conselho Nacional de Pesquisas (National Research Council) was created and would have a crucial role in the institutionalization and expansion of scientific activity.[52] In a way, the council was the consolidation of the efforts of a small group of researchers from the Academia Brasileira de Ciências (Brazilian Academy of Science), which, since the 1920s, had promoted the need for governmental support and financing of scientific research.

During the 1950s, a small number of elite teaching and research institutions were established; they served as model and inspiration for the broadest reforms in the system of universal education that would take place next. The first of these elite institutions was the Instituto Tecnológico de Aeronáutica (Technological Aeronautics Institute, ITA), sponsored by the Massachusetts Institute of Technology; it carried out research and educated military personnel in how to engage in R&D. The ITA changed the university curriculum by stressing the need for a scientific basis for technological applications with an important experimental element.

The Faculdade de Medicina (Medical School) of Riberão Preto was also important. It was there that the most modern biomedical research model was developed and academic departments were organized according to new disciplines derived from recent contributions from physics to the biological sciences. The Rockefeller Foundation's input—which in the previous period had supported Houssay's Instituto de Fisiología in Argentina—was crucial during this period for the school's early development and introduction.

The third important institutional experience during this period was the creation of the Universidade de Brasília, an ambitious, revolutionary, and imaginative project that did not have time to mature because it quickly fell victim to ideological conflicts and military repression in 1964. Although it kept its reputation as a good federal university, it never recovered its initial mystique and prestige.

In Mexico, scientific research was closely related to the Facultad de Ciencias (Science Department) of the Universidad Nacional Autónoma de México (National Autonomous University of Mexico, UNAM) and to measures intended to legitimate the role of the full-time researcher within the university during the 1940s. In 1950, when the University City was begun, the Facultad de Ciencias building was the first to be constructed; shortly thereafter, the Torre de Ciencias (Science Tower) was begun. For the first time, a specific space was assigned to scientific research. Nonetheless, research budgets remained small, although some persons were being hired to do research exclusively, in spite of the lack of support for experimental work.[53]

The Age of Scientific Policy, 1960–1980

It was during the 1960–1980 period that economic planning organizations began to operate. It is evident in their first development reports that they were limited by the lack of institutional coordination; short-, long-, and middle-term inconsistency; and the absence of adequate personnel, projects, and statistics. Sometimes planning itself emerged as a mechanism for obtaining funding from international organizations. The background context was one of contradictory developments. The prevailing ideology of modernization was expected to produce greater autonomy, self-confidence, and social justice. The social dynamic of this ideology found widespread expression, and not only in scientific activity. The 1960s in particular was a period of increasing self-confidence, optimism, and hope for building more just and equal societies. The scientists, engineers, and governmental functionaries who tried to apply their projects to practical ends, as, for example, atomic energy in Argentina and electronics in Brazil,[54] also saw some success. For a while, they managed, with their unexpected contributions, to change the rules of the game. The development of local capacity in science, technology, industry, management, and labor-force skills introduced significant changes in local social structures. It also created a new set of actors and provided them with a better understanding of the art of negotiation.

But these changes were not enough to alter the social and economic conditions that ultimately caused the effort to fail: a pattern of economic development based on growth without social equality;[55] internal market-oriented industrialization, biased to favor the conspicuous consumption of luxury goods at levels that were significantly higher than in other countries with delayed industrialization and comparable income levels;[56] lack of leadership in private companies and within the most dynamic industrial sectors (automobile manufacturing, chemistry, and capital-goods production), which are the vectors of technological progress and which shape the profile of national productivity; weak influence of small and medium-sized industry; limited participation of the national private sector in R&D, even in the most advanced countries of the region;[57] distorted and underdeveloped enterprise capabilities; inadequate growth rates; profound regional imbalance; a marked concentration of income; growing globalization; and substantial increases in debt. Highly correlated to these situations, authoritarian regimes began to emerge: in Brazil in 1964; Peru, 1968; Ecuador, 1969; Bolivia and Uruguay, 1970; Chile, 1973; and Argentina, 1974.

During this period, there were ambitious attempts to radically change traditional university structures and to give scientific and technological research a more central role in economic and social planning. Scientific and technological research outside of the university was strongly encouraged in both the public and the private sectors and by the institutions dedicated to basic or applied research. Institutions of higher education increasingly began to adopt the U.S. model of centralized institutes and departmental organization. Postgraduate education began to appear as a regular component of university programs. More full-time teaching jobs were made available in universities. At the same time, university entrance requirements were lowered. In sum, the system of higher education became much larger, and the rapid growth in the number of students became probably the most crucial problem for universities, whose budgets were never large enough to respond to the explosion in demand. The science and technology councils began to finance research that could not be undertaken at academic institutions, and they have tried since then to define priorities and to reorient scientific activity.[58]

By the late 1960s, the Brazilian government had made the greatest effort to link scientific development and economic development. Its efforts were innovative, for the financing of scientific and technological development came from the government, which was responsible for economic planning and long-term investment. This meant that the money available for research surpassed the country's research capability and that short-term efficiency and productivity criteria were frequently used to evaluate research activity. This coincided with profound transformations in the country's system of higher education, for differentiation and stratification deepened, allowing a parallel system of private institutions to develop to compensate for the limited number of spaces at public universities.

Venezuela's rapid modernization during this time also stands out. It began before the period in question but increased in the 1960s and the 1970s, helped by the immense increase in income from oil. Although the Asociación Venezolana para el Avance de la Ciencia (Venezuelan Association for the Advancement of Science) was founded in 1950 by a small group of scientists, and in 1952, the Fundación Luis Roche—a true seedbed for modern scientific research—was created, the dictatorship of Marcos Pérez Jiménez had to fall in 1958 before science could begin to institutionalize as society democratized.[59]

The educational system was stimulated at all levels. The Universidad Central de Venezuela (Central University, UCV) began systematically to consider science as one of its educational goals. A law was enacted emphasizing research as one of the university's basic functions and activities. The Facultad de Ciencia was established at the UCV, and the Consejo de Desarrollo Científico y Humanístico (Council for Humanistic and Scientific Development) was organized to influence positively the university environment by encouraging research.[60] With the idea of complementary functions, the Instituto Venezolano de Investigaciones Científicas (Venezuelan Institute of Scientific Research, IVIC) was created as an excellence center with the exclusive purpose of carrying on full-time scientific research. Although later the IVIC also offered postgraduate courses, it is still considered a pure research center, with features unique in Latin America.[61]

During the 1960s, foundations and universities from the United States frequently sent experts and established cooperative programs, much as had happened in Brazil during the 1950s. At this time, an abrupt switch from the old teaching methods to modern academic methods took place. Courses were modified and new classes were introduced. The country's new oil wealth allowed the hiring of foreign researchers to institute new avenues of research locally. As in other Latin American countries (such as Argentina and Brazil), the 1960s ended with the 1968 renovation movement. In Venezuela, the movement—started by the Facultad de Ciencias—sought to redefine the university in order to train ideologically mature and academically serious scientists and engineers who, with their transformative political perspective, would be incorporated into society. The movement ended in failure, and in 1970, the university was taken over and restrictive regulations were imposed.

The Consejo Nacional de Investigaciones Científicas y Técnicas (National Council of Scientific and Technical Research, founded in 1967) became, in the mid-1970s, the voice of a new group of technicians and social scientists who demanded the development of national technological capability. This should not be confused with the development of policies for or from science. But such efforts did not begin to gain strength until the 1980s.

During the university post-transformation period, in 1970, two universities emerged: the Universidad Simón Bolívar, financed by the Inter-American Development Bank; and the Universidad Metropolitana, supported by local business groups. Both were markedly oriented toward engineering; the expansion of the sciences continued at the autonomous universities.[62]

One of the most characteristic phenomena of this period was the appearance of public and private research institutions as a response to the multiple pressures (from students, the right wing, the military, the church) that threatened researchers' well-being and continuity at the university, the almost exclusive institutional context for research to that point. The major U.S. foundations collaborated in the creation of a private research sector. This was particularly evident in the social sciences as authoritarian governments withdrew all support.[63] In several Latin American countries, the state preferred to remove research from universities, especially research that was oriented toward specific purposes, rather than transfer it to public institutions and companies, as happened in Brazil.

Some efforts emerged from the civil sector as well. These included some connected to the private sector, others organized for the explicit purpose of carrying out disinterested theoretical research in a particular scientific discipline independently of the government. Finally, some private foundations, both national and foreign, emerged.[64] The new groups that benefited from the growing funds available for scientific and technological research tended to be young and apolitical or, at least, had a short memory and few personal ties to the recent past. Working in quite isolated and protected places, earning salaries—independent from university budgets—and without having to teach undergraduate students, they frequently thought of themselves as long-term reformers who were waiting for the political storm to pass to set up the foundations of scientific and technological confidence in their countries.[65]

A New Audience for Science:
Industrial Companies, 1980–1990

The Latin American industrial sector during the 1980–1990 decade has been described as having the following features: relatively high margins of idle capacity in several countries and sectors and precarious finances in associated firms along with a decaying internal market; excessive debt and high interest rates; drastically reduced investment rates in several countries, which, in a time of rapid technological change in the capital goods sector worldwide, increased technological obsolescence; and, in some cases, weak or defunct design groups in factories and engineering firms and lowered qualifications for the industrial labor force, which transferred to other activities when workers were fired.[66]

As for the public sector, the combined effects of limited investment, concentration on short-term problems and the consequent negligence of strategic thinking, and drastic wage shrinkage weakened support to certain crucial areas such as R&D. The state's economic limitations were also evident in the crisis in education at all levels. Thus, what is in question today is not a temporary difficulty but the entire productive and social system.

The traditional and well-established universities, which historically fostered research groups, decayed. The small research community had to compete against an increasingly larger sphere of university teachers, who had full-time, stable jobs but who did not conduct research. The universities lost their appeal as a privileged locus for research. Scientists and engineers tried, whenever possible, to organize their work outside academia or around isolated postgraduate programs. Thus, the mortality rate and diversity of objectives, as well as the obvious differences in quality, guaranteed that the old mechanism of granting scholarships to students in order to send them to centers in the developed countries would continue as long as there were funds and opportunities.[67]

The possibilities of change in and modernization of universities on a global scale seem remote. Only a few institutions concern themselves intensely with scientific research and educating future researchers. This does not mean, however, that the research that is being conducted at public universities should be discounted. Some of the scientists who decided to remain in academia looked—beyond traditional governmental support—for a public that appreciated them for their ability to educate, to innovate, and to offer expert advice more than for their publications and scientific renown—although they relied on their initial prestige as a way of replacing diminished governmental support. The rhetoric of industrial utility finally arrived in Latin America with all the strength of the 1980s, but that rhetoric collided with two difficulties. On the one hand, the opportunities for an industrial science and a highly qualified labor force were not great; on the other, a dangerous abyss was opened between what was supposedly "useful" or at least "saleable" and what was purely cognitive. New groups began to compete for scarce resources with qualified or more-established scientific groups. When the traditional mechanism of peer review operates, the most competitive groups have a better chance to prevail in the struggle for scarce resources. When participants with new criteria or support sources enter the decision-making process, the situation can be reversed. Sometimes it is just a matter of pumping new life into old evaluation and decision-making mechanisms. Very frequently, it is a simple matter of conflict between competence on one hand, and incompetence and intellectual opportunism on the other.

During the 1970s, there were few places where the industrial alternative to private support was realistic. In contrast, during the 1980s, a new alliance between academic science and utility began to develop and to spread to several areas of learning, beyond the traditional ones. The fields that are most involved are, however—and as expected—agricultural engineering, biotechnology, veterinary medicine, pharmacy, the exact sciences, and administration.

The case of the exact sciences is particularly interesting. We might assume that, given the emphasis on basic research, this area was the least desirable with which to establish cooperative agreements, since such agreements are closely related to applied research and the development of experimental research and technical assistance. Nevertheless, there were in several countries institutions in which the exact sciences could be found to be working closely with the private sector, a cooperation that sometimes was greater than that with the different engineering branches. It was at those institutions that some very innovative solutions were tested. Sometimes this cooperation also implied an increase in experimental physics to the detriment of theoretical physics; in other cases, a division of functions was adopted, with the university assuming basic research and industry taking on applied research. That is to say that, in this case, the universities cooperated with those companies that had their own research laboratories.[68]

Undoubtedly, the large public enterprises were the main clients of the universities. This preference certainly was not the result of any political or ideological choice but the simple consequence of the fact that in countries such as Argentina, Brazil, and Mexico, the primary state-owned enterprises represented the most important part of the productive sector. At the same time, they encompassed the most advanced component of the economy, and they were the only companies that had the resources for such cooperative relationships, as well as the awareness of the strategic importance to the economy of the link between academia and the productive sector.

Nevertheless, the dominance of the large state-owned enterprises does not mean the private sector was absent, especially those companies that used more advanced technology. Here, the primacy of the large national or multinational companies can be observed again. The multinationals with interests and operations in Latin America deserve special attention.

For a long time, it was assumed that the collaboration between these companies and the local universities could only be limited, since the technologies and processes used were presented as imported "packages," the result of research and products developed in Europe and the United States. This fact, indisputable until recently, seems increasingly doubtful and difficult to justify. Actually, many products and technologies need to be adapted to the Latin American market and conditions, which implies a considerable research and technological development effort on the part of Latin American universities, which are certainly better qualified for this than their European or U.S. counterparts. In addition, Latin America offers foreign companies a particular advantage: significantly lower R&D costs.

The trend is that the relationship between these two environments—the academic and the productive—will strengthen. Some sectors enjoy a dynamism and motivation that are, on one hand, the consequence of the financial crisis of the universities, which demands alternative financing. On the other, they result from the relationship between the university and industry. We can conclude that these cooperative efforts might be strategic for these countries. The university represents the scientific spirit and society's emergent knowledge and the world of economy and work, where productive capacity and quality are developed. Both of these exist in a world in which activity is increasingly internationalized and in which markets are progressively more open and more competitive at the same time.

Conclusions

This superficial recounting of scientific developments in twentieth-century Latin America shows, among other things, that the new scientific community was created in permanent counterpoint to the will to incorporate the international scientific system and the desire to have a voice of its own, autonomy in defining its profile, interests, and legitimization. Its position on the periphery seems to have become more acute in the current political, economic, and social crisis. Much of the freedom and autonomy a nation has to make decisions comes from its scientific and technological capabilities. Knowledge means, now more than ever, power and opportunity.

Latin America, though, has not yet developed an enduring consensus regarding this axiom. The policies regarding science and technology have been like stop-and-go accordion playing: sometimes there are not enough researchers to respond to ambitious governmental programs; at other times, there is not enough money to support the existing research capacity in a country. The most recent studies point to a serious worsening of the working conditions within the scientific environment and to the growing alienation of researchers, who frequently lack support for even minimal needs for their work. Also, a number of critics point out that much of what is done is trivial and that "applied research ideology"—which is what we call the developments of the 1990s—could have helped strengthen research capacity. At the same time, the international situation continues its dynamic movement, which has reduced the space available to Latin America. The current process of internationalization of the economic system is openly favorable to the more industrialized countries.

The individuals and groups that, from a wish for self-determination and scientific and technological training, defended the development of local productive forces in Latin America find themselves in a defensive position at the present time, or are openly ignored as obsolete. The development of local scientific capacity in the region is discouraged in many ways. But the solution, no matter how stubborn it might seem, appears to be as valid today as during the 1960s: Latin America must guarantee the existence and expansion of local research capacity as a condition necessary—although not the only one—to success. Success, ultimately, will depend on radical social transformations and careful international negotiations.

Notes

1. I subscribe to Octávio Ianni's idea of analyzing how Latin American thinking was formed: as a history of the idea of Latin America. Mine is an effort to complement him by including science within the range of the history of Latin American cultural thinking (Ianni, "A idéia").

2. The essence of scientific institutions lies in the achievements of thoughts that are formulated and communicated as institutional ideas, roles, and functions (Adler, *The Power*).

3. When I consider the institutionalization of modern science in twentieth-century Latin America, I do not assume it to be the starting point for science in the region. There existed several scientific, technological, or artisanal activities even before the Conquest and occupation, and they underwent diverse changes.

4. When adopting positivism, though, Latin American intellectuals made some changes to Comte's thinking. The phrase "Love, order and progress" was changed by the Mexican Liberal Party into "Liberty, order and progress." The "liberty" component was inserted only to articulate the party's program and the reality of the established regime (De Gortari, *La ciencia*, p. 81). Soon, liberty disappeared from Mexican positivist thinking except in those cases in which it was considered destructive and therefore had to be condemned. Brazil's motto was "Order and progress," where order was the existing order, presented as possibly the best in "realist" conditions. As for progress, it was stated that it could be attained only within the established order. It was presented as a gradual evolution from which the possibility of revolution was excluded. Thus, progress was reduced to public works and to the growth of the bourgeoisie's wealth.

5. The classical contributions of taxonomic sciences date from the nineteenth century: descriptions of varying levels of detail and precision; observations on the subcontinental flora, fauna, and minerals; meteorological phenomena; topographic features and other geographic accidents and their geological constitution; astronomical data; the recording of clinical medical practice and ethnographic descriptions.

6. For further information about these men, see De Gortari (*La ciencia*) for Mexico; Babini (*El pensamiento*) for Argentina; and Schwartzman (*Formação*) for Brazil.

7. See Glick, "Perspectivas," p. 50.

8. Safford, "Acerca de la incorporación," p. 426.

9. The controversy originated mainly from Burmeister's resistance to acknowledging Ameghino's merits. Burmeister held for a long period, a position of power as director of the Museo de Historia Natural (Natural History Museum) of Buenos Aires and kept Ameghino from developing a career in science locally,

even long after he was recognized in European circles and even after Burmeister's death.

10. By this time, Ameghino had published two works that demonstrated his maturity: *La formación pampeana* (The Formation of the Pampas) and *La antigüedad del hombre en el Plata* (The Antiquity of Man in El Plata). In 1882, Ameghino marshaled his opinions about transformism in a paper entitled "A la memoria de Darwin" (To the Memory of Darwin), and two years later did the same thing regarding his opinions about evolution in *Filogenia* (phylogeny), the content of which he defined as "classificatory transformist principles based on natural laws and mathematical proportions" (Romero, *El desarrollo*).

11. Glick, "Perspectivas."

12. Cueto, "Nacionalismo," pp. 327–355.

13. Cueto, *Excelencia*.

14. To name only a few such institutions: the Facultades de Medicina y Ciencias (Schools of Medicine and Science) of the Universidad de San Marcos, the Escuela de Ingenieros (Engineering School) and the Escuela de Agricultura (Agriculture School) in Peru (ibid.; Sociedad Peruana de Historia de la Ciencia y la Tecnología [SOPHYCIT], *Estudios*); the Observatorio Astronómico Nacional (National Astronomical Observatory) (Moreno, "El Observatorio Astronómico Nacional") and the Academia Nacional de Ciencias y Literatura (National Academy of Sciences and Literature) in Mexico; the Escola de Minas (School of Mines) of Ouro Preto (Carvalho, *A Escola de Minas*), the Museu de História Natural (Natural History Museum), the Observatorio (Morize, *Observatório*), and the Instituto de Manguinhos in Rio de Janeiro, Brazil.

15. Engineers trained at São Paulo's Escola Politécnica (Polytechnic School, founded in 1874), the Escola de Minas of Ouro Preto (established in 1876), the Escola Politécnica of São Paulo (1893), and in four other engineering schools in other cities founded soon after these (Telles, *História*, pp. 48–62).

16. Safford, *The Ideal of the Practical*, p. 225.

17. "Colombian engineers in fact did little of note in the way of innovation. They took little interest in tropical agriculture, a field in which they might have made some contribution. And in other areas the country's lagging economy posed few problems for which some kind of solution could not be found in the more advanced countries" (ibid., p. 226).

18. Stepan, *Beginnings*; Benchimol, *Manguinhos*.

19. Babini, *El pensamiento*, p. 110.

20. A typical example of the transition from amateur activity to professional and scientific activity in Brazil is the laboratory the Osório Almeida brothers built in their house in Rio. It was one of the most important centers for training physiology students in the country. Much research remains to be done regarding these brothers.

21. Babini, *El pensamiento*, p. 109.

22. This does not mean nothing existed before. For a treatment of science's role in the birth of the Mexican state, see Saldaña, "La ciencia," pp. 37–52.

23. Examples include the agricultural experiment stations in Argentina, the Instituto Agronómico (Agronomy Institute) of Campinas, Brazil, the modernization of the cities, harbors, and so on, which are impressive symbols of progress

and industrialization. In this spirit, the urban reform of Rio de Janeiro, carried on by engineers determined to "regenerate" the city, involved the destruction of the most obvious signs of the old city.

24. Costa Ribeiro, "Acesso."

25. We must also remember that by the end of the nineteenth century similar developments had occurred in advanced countries such as France, where the creation of true universities, instead of the mere juxtaposition of departments, was discussed, but had to wait until 1968 for the universities to take control of the departments (Weisz, *The Emergence*).

26. The first eruption of the "university reform" movement occurred in Córdoba, at whose university bizarre colonial traits persisted and where the traditional oligarchy and the clergy were stronger than in Buenos Aires. Socialist senator Juan B. Justo made a vigorous speech to Congress on July 24, 1918, stating that what was needed was a total housecleaning, throwing "if not through the window at least through the door all that is apocryphal in the Universidad de Córdoba, all apocryphal science, all verbal and phony science, all negligent, ignorant, and inept professors" (Palacios, *La universidad nueva*).

27. In 1930, a brief reformist experiment was carried out at the Universidad de San Marcos. Administrators tried to put into practice the principles of the 1918 university reform but ended the experiment in 1932, after violent repression by the military regime, which closed the university for more than three years.

28. The study of geology began in 1876 as a part of the mining engineering program at the old Escuela de Ingenieros, now the Universidad Nacional de Ingeniería (National Engineering University), but its denationalization in the first decade of the twentieth century made mining engineering unattractive to Peruvians.

29. Podestá and Olson, "Predicción," p. 37.

30. Safford, *The Ideal of the Practical*, pp. 231–242.

31. Trujillo Cenóz and Macadar, "Biología."

32. Petitjean, Jami, and, Moulin (eds.), *Science and Empires*.

33. In 1921, the Instituto Francés (French Institute) of Buenos Aires (later known as the Institut de l'Université de Paris à Buenos Aires), was created, but the political interests of the French authorities did not meet Argentina's needs and money for this institute was scarce. In 1922, the Institut Franco-Bresilien de Haut Culture (Franco-Brazilian Institute for High Culture) was founded, as well as a Department of French Culture in Argentina and Chile. In 1924, the Instituto Francés Mexicano (Franco-Mexican Institute) was created, and similar institutions were founded in Lima in 1927 and in Caracas in 1928.

34. Cueto, "The Rockefeller Foundation."

35. Brunner and Barrios, *Inquisición*.

36. Pyenson, "*In Partibus Infidelium*."

37. Because of his gift as researcher and teacher, Bose quickly became unpaid lecturer and assistant to Nernst. In 1906, he was named director of the Technische Hochschule (Institute of Technology) in Danzig as a strategy to "Germanize" those parts of Prussia that were Polish speaking.

38. Roca Rosell and Sánchez Ron, *Esteban Terradas*, pp. 217–260.

39. Among the first holders of the chair were Ramón Menéndez Pidal (1914), José Ortega y Gasset (1916), Julio Rey Pastor (1917), Augusto Pi i Sunyer

(1919), Blás Cabrera (1920), María de Maeztu (1926), and Esteban Terradas (1927); other Spanish intellectuals invited by other institutions held the position of visiting professor (Gutiérrez, in Roca Rosell and Sánchez Ron [eds.], *Esteban Terradas*).

40. See Kenny et al., *Inmigrantes*, for Spanish immigrants and refugees in Mexico. Roca Rosell and Sánchez Ron's book about Terradas (*Esteban Terradas*), the "shipwreck survivor of the civil war in Argentina," and his fruitful work at the Facultad de Ingeniería at the Universidad de Buenos Aires, particularly in aeronautics, is interesting.

41. For an introduction to Prebisch's work, see CEPAL, *Raúl Prebisch*.

42. Chudnosky et al., *Capital Goods*.

43. Klubitschko, *Posgrado*.

44. Antonorsi and Ávalos, *La planificación ilusoria*.

45. Katz, *Oligopio*.

46. Fajnzylber, *La industrialización*.

47. Vessuri, "El proceso."

48. It is hardly a surprise that most of the debate surrounding national science policy in Latin America was centered around the role of the university, particularly its research function and its role as a provider of highly qualified personnel. After all, in most countries, the universities encompassed a substantial portion of the existing research capacity.

49. In Brazil, the institutions of higher education founded at the beginning of the nineteenth century, when the Portuguese court was transferred to Brazil, were oriented toward the education of liberal professionals; that orientation remained unchanged throughout the century. The strength of the positivist movement blocked the establishment of universities. Brazilian positivists objected to the Napoleonic university model, thus ignoring the transformations the institution had undergone in the nineteenth century, especially in Germany.

50. Among the first professors were Luigi Fantappié (mathematics) (Hönig and Gomide, "Ciências matemáticas," vol. 1, p. 45), Gleb Wataghin (physics) (Motoyama, "A física no Brasil," vol. 1, pp. 73ff), Heinrich Rheinboldt (chemistry) (S. Mathias, "Evolução da química no Brasil," vol. 1, p. 103), Felix Razawitscher (botany) (M. Guimãraes Ferri, "História da botánica no Brasil," vol. 2, p. 66), Ettore Onorato (mineralogy) (R. Ribeiro Franco, "A mineralogia," vol. 3, pp. 34–36).

51. All of the society's main organizers were affiliated with biological research institutions in São Paulo, thus reflecting the concentration of research activity in that city (Schwartzman, "A Space," p. 249).

52. At the end of the 1970s, it changed its name to Conselho Nacional de Desenvolvimento Científico e Tecnológico (National Scientific Development and Technology Council).

53. "Pena."

54. Adler, *The Power*.

55. Fajnzylber, *La industrialización*.

56. Esser, "La inserción."

57. Katz, *Desarrollo*.

58. The mostly illusory character of this strategy is analyzed by Antonorsi and Ávalos in a book about scientific and technological planning in Venezuela that has become a classic, *La planificación ilusoria.*
59. Roche, "El discreto encanto."
60. Vessuri, "El proceso."
61. Vessuri, "El papel."
62. Vessuri, "La formación."
63. Brunner and Barrios, *Inquisición;* Vessuri, "El Sísifo."
64. We have not been able to study research institutions tied to private firms because of difficulties related to the material.
65. Schwartzman, "The Quest."
66. Fajnzylber, *La industrialización.*
67. Brazil and Mexico have developed postgraduate education more than the other countries of the region, although some fourth-level programs have been established in almost all the countries of the region.
68. Grilo, Cerych, and Vessuri, "As relações."

Bibliography

Adler, E. *The Power of Ideology: The Quest for Technological Autonomy in Argentina and Brazil.* Berkeley & Los Angeles: University of California Press, 1987.
Antonorsi, M., and Ávalos, I. *La planificación ilusoria.* Caracas: CENDES/Ateneo, 1980.
Babini, J. *El pensamiento científico en la Argentina.* Buenos Aires: La Fragua, 1954.
Benchimol, J. L. (coord.). *Manguinhos do sonho á vida: A ciência na Belle Époque.* Rio de Janeiro: Casa de Oswaldo Cruz, 1990.
Brockway, L. *Science and Colonial Expansion: The Role of the British Royal Botanic Gardens.* New York: Academic Press, 1979.
Brunner, J. J., and A. Barrios. *Inquisición, mercado y filantropía: Ciencias sociales y autoritarismo en Argentina, Brasil, Chile y Uruguay.* Santiago: FLACSO, 1987.
Carvalho, J. M. de. *A Escola de Minas de Ouro Preto: O peso da glória.* Rio de Janeiro: Financiadora de Estudos y Projetos/Editora Nacional, 1978.
Comisión Económica para América Latina y el Caribe (CEPAL). *Raúl Prebisch: Un aporte al estudio de su pensamiento.* Santiago, 1987.
Chudnosky, D., et al. *Capital Goods Production in the Third World: An Economic Study of Technology Acquisition.* London: F. Pinter, 1983.
Costa Ribero, S. "Acesso ao ensino superior: Uma visão." Paper presented at "Situation and Perspectives of Superior Education in Brasil" conference. São Paulo: Núcleo de Pesquisas sobre Ensino Superior/Universidade de São Paulo, 1989.
Cueto, M. *Excelencia científica en la periferia: Actividades científicas e investigación biomédica en el Perú, 1890–1950.* Lima: GRADE/CONCYTEC, 1989.
———. "Nacionalismo y ciencias médicas: Los inicios de la investigación biomédica en el Perú: 1900–1950." *Quipu, Revista Latinoamericana de Historia de las Ciencias y la Tecnología* 4, no. 3 (September–December 1987): 327–356.

———. "The Rockefeller Foundation and the Latin American Physiology." *Research Reports from the Rockefeller Archive Center*. North Tarrytown, N.Y., 1990.
De Gortari, E. *La ciencia en la historia de México*. Mexico City: Grijalbo, 1973.
Esser, K. "La inserción de América Latina en la economía mundial: Integración pasiva o activa." *Integración Latinoamericana* 12, no. 126 (1987): 35–57.
Fajnzylber, F. *La industrialización trunca de América Latina*. Mexico City: Nueva Imagen, 1983.
Franco, R. Ribeiro. "A mineralogia e a petrologia no Brasil." In *História das ciências no Brasil*, M. Guimarães Ferri and S. Motoyama (coords.). São Paulo: EDUSP, 1979–1981.
Ferri, M. Guimãraes. "História da botánica no Brasil." In *História das ciências no Brasil*, M. Guimarães Ferri and S. Motoyama (coords.). São Paulo: EDUSP, 1979–1981.
Ferri, M. Guimãraes, and S. Motoyama (coords.). *História das ciências no Brasil*. 3 vols. São Paulo: EDUSP, 1979–1981.
Glick, T. "Perspectivas sobre la recepción del darwinismo en el mundo hispano." In *Actas II Congreso de la Sociedad Española de Historia de las Ciencias*, M. Hormigón (ed.). Jaca: Sociedad Española de Historia de las Ciencias, 1982.
Grilo, E. M. L. Cerych, and H. Vessuri. "As relações universidade–sector productivo nos países da América Latina." RIEIE 3, no. 1 (1990): 12–29.
Hönig, S., and E. F. Gomide. "Ciências matemáticas." In *História das ciências no Brasil*, M. Guimarães Ferri and S. Motoyama (coords.). São Paulo: EDUSP, 1979–1981.
Ianni, O. "A idéia de América Latina." Draft, no. 13 (1990). Instituto de Filosofia e Ciências Humanas/Universidade Estadual de Campinas.
Katz, J. *Desarrollo y crisis de la capacidad tecnológica latinoamericana: El caso de la industria metalmecánica*. Buenos Aires: Comisión Económica para América Latina, 1986.
———. *Oligopolio, firmas nacionales y empresas multinacionales: La industria farmacéutica argentina*. Buenos Aires: Siglo XXI, 1974.
Kenny, M., et al. *Inmigrantes y refugiados españoles en México: Siglo XX*. Mexico City: Casa Chata, 1979.
Klubitschko, D. *Posgrado en América Latina*. Caracas: Centro Regional de la UNESCO para la Educación Superior en América Latina y el Caribe, 1986.
Mathias, S. "Evolução da química no Brasil." In *História das ciências no Brasil*, M. Guimarães Ferri and S. Motoyama (coords.). São Paulo: EDUSP, 1979–1981.
Moreno, M. A. "El Observatorio Astronómico Nacional y el desarrollo de la ciencia en México (1878–1910)." *Quipu, Revista Latinoamericana de Historia de las Ciencias y la Tecnología* 5, no. 1 (1988): 59–67.
Morize, H. *Observatório astronómico: Um século de história (1827–1927)*. Rio de Janeiro: MAAST, 1987.
Motoyama, S. "A física no Brasil." In *História das ciências no Brasil*, M. Guimarães Ferri and S. Motoyama (coords.). São Paulo: EDUSP, 1979–1981.
Myers, J. "Antecedentes de la conformación del sector científico y tecnológico, 1850–1958." In *Examen de la política científica y tecnológica nacional*, E. Oteiza

(coord.). *Perspectivas a Mediano Plazo* series. Buenos Aires: Secretaría de Ciencia y de Tecnología/Programa de las Naciones Unidas, 1989.

Palacios, A. *La universidad nueva: Desde la reforma universitaria hasta 1957*, M. Gleizer (ed.). Buenos Aires, 1957.

"Pena, estado actual y perspectivas de las ciencias biológicas en México." In *La biología como instrumento de desarrollo para América Latina*, J. E. Allende (ed.). Santiago: Red Latinoamericana de Ciencias Biológicas, 1990.

Petitjean, P.; C. Jami; and A. M. Moulin (eds.). *Science and Empires: Historical Studies about Scientific Development and European Expansion*. Boston Studies in the Philosophy of Science 136. Boston: Kluwer Academic Publishers, 1992.

Podestá, B., and R. S. Olson. "Predicción de un sismo, su trama y manejo." *Quipu, Revista Latinoamericana de Historia de las Ciencias y la Tecnología* 4, no. 1 (January–April 1987): 33–52.

Pyenson, L. "*In Partibus Infidelium*: Imperialist Rivalries and Exact Sciences in Early Twentieth-century Argentina." *Quipu, Revista Latinoamericana de Historia de las Ciencias y la Tecnología* 1, no. 2 (1984): 253–303.

Ríos, S. L. A. Santaló, and M. Balanzat. *Julio Rey Pastor, matemático*. Madrid: Instituto de España, 1979.

Roca Rosell, A., and J. M. Sánchez Ron (eds.). *Esteban Terradas: Ciencia y técnica en la España contemporánea*. Madrid: International Trademark Association/Serbal, 1990.

Roche, M. "El discreto encanto de la marginalidad: La Fundación Luis Roche." In *Las instituciones en la historia de la ciencia en Venezuela*, H. Vessuri (coord.). Caracas: Fondo Editorial Acta Científica Venezolana, 1987.

Romero, J. L. *El desarrollo de las ideas en la sociedad argentina del siglo XX*. Buenos Aires: Fondo de Cultura Económica, 1965.

Safford, F. *The Ideal of the Practical: Colombia's Struggle to Form a Technical Elite*. Austin: University of Texas Press, 1976.

———. "Acerca de la incorporación de las ciencias naturales en la periferia: El caso de Colombia en el siglo XX." *Quipu, Revista Latinoamericana de Historia de las Ciencias y la Tecnología* 2, no. 3 (September–December 1985): 423–435.

Saldaña, J. J. "La ciencia y el Leviatán mexicano." *Actas de la Sociedad Mexicana de Historia de la Ciencia y la Tecnología* 1 (1989): 37–52.

Schiefelbein, E. "Equality Aspects of Higher Education." Unpublished. Washington, D.C., 1985.

Schwartzman, S. *Formação da comunidade científica no Brasil*. Rio de Janeiro: Editora Nacional, 1979.

———. "The Quest for University Research: Policies and Research Organization in Latin America." In *The University Research System: The Public Policies of the Home of Scientists*, B. Wittrock and A. Elzinga (eds.). Stockholm: Almqvist and Wiksell International, 1985.

———. "A Space for Science: The Development of the Scientific Community in Brazil." Unpublished, 1988.

Sociedad Peruana de Historia de la Ciencia y la Tecnología. *Estudios de la historia de la ciencia en el Perú*. 2 vols. Lima: Consejo Nacional de Ciencia y Tecnología, 1986.

Stepan, N. *The Beginnings of Brazilian Science: Oswaldo Cruz, Medical Research and Policy, 1890–1920*. New York: Columbia University Press, 1976.
Telles, P. C. da Silva. *História da engenharia no Brasil. Anais, I Seminario Nacional sobre História da Ciencia e Tecnologia*. Rio de Janeiro: Museu da Astronomia e Ciências Afins/Conselho Nacional de Desenvolvimento Científico e Tecnológico, 1986.
Trujillo Cenoz, O., and O. Macador. "Biología." In *Ciencia y tecnología en el Uruguay*. Montevideo: Ministerio de Educación y Cultura, 1986.
Tunnermann Bernheim, C. *60 años de la reforma universitaria de Córdoba, 1918–1978*. Caracas: Fondo Editorial para el Desarrollo de la Educación Superior, 1979.
Vessuri, H. M. C. "La formación de la comunidad científica en Venezuela." In *Ciencia académica en la Venezuela moderna*, H. Vessuri (coord.). Caracas: Fondo Editorial Acta Científica Venezolana, 1984.
———. "El papel cambiante de la investigación científica académica en un país periférico." In *La ciencia periférica: Ciencia y sociedad en Venezuela*, E. Diáz, Y. Texera, and H. Vessuri (coords.). Caracas: Monte Ávila, 1984.
———. "El proceso de profesionalización de la ciencia venezolana: La Facultad de Ciencias de la Universidad Central de Venezuela." *Quipu, Revista Latinoamericana de Historia de las Ciencias y la Tecnología* 4, no. 2 (May–August 1987): 253–281.
———. "El Sísifo sureño: Las ciencias sociales en la Argentina." *Quipu, Revista Latinoamericana de Historia de las Ciencias y la Tecnología* 7, no. 2 (1990): 149–185.
Weisz, G. *The Emergence of Modern Universities in France, 1863–1914*. Princeton, N.J.: Princeton University Press, 1983.

CHAPTER 8

Excellence in Twentieth-century Biomedical Science

MARCOS CUETO

For a long time, the development of science in Latin America was considered a poor imitation of the history of scientific development in the industrialized countries. This idea deprived the region's science of a dynamic past of its own and guided the actions of the North American philanthropic agencies for many years. It inspired the first efforts toward modernization, that proposed a linear science-development model.[1] This model assumed that the development of science would go through the same phases in all countries.

The first critical studies of Latin American science that reacted to these models emphasized the concept of "periphery."[2] This concept was (and probably still is) useful for locating the scientific communities that were considered to be on the "outskirts" of the traditional centers of knowledge.

We can justifiably raise a number of objections to this concept's true theoretical origin: dependency theory. One of the most important objections concerns the lack of temporality, the linearity, and the passivity assigned to the periphery. The periphery concept tends to overlook the importance of many Latin American researchers who worked at the international level without ignoring the social and cultural influence of their own locality. It also focuses researchers' attention exclusively on failures and on explaining something that, in fact, never took place. This concept assumes that, had it not been for some kind of contact between the native scientific communities and those in the developed countries, development would have been different. This kind of reasoning is difficult to demonstrate from the historical point of view, especially when we consider that explaining what actually occurred is the historian's primary duty.

Even though I acknowledge the term "periphery" as a useful spatial reference, I think its meaning should be revised, not only because it prevents a clear understanding of important instances of scientific excellence but also because it minimizes the role played by Latin American scientists in the construction of their own history. The studies that emphasize the periphery concept do not work well when they have to explain the dynamics of the region's scientific development. It is only recently that developing countries have begun to sketch the rules inherent in the history of science. Even when applied to cases of scientific excellence, such rules are somehow different from those used to build central institutions in the industrialized countries.

In this short essay, I intend to present and to analyze the main components of one of the most significant developments in Latin American biomedical science: the effect of Peru's altitude on its citizens' physiology. In addition, I will try to establish comparisons with other successful episodes in the study of Latin American physiology and microbiology in the early decades of the twentieth century. One of my aims is to contrast these components with the assumptions that guided the actions of North American philanthropic agencies and that inspired the linear models of scientific modernization. First, scientific development occurs in isolation from other cultural or social changes. Science was presented in this assumption as a form of knowledge that should develop unrelated to any kind of foreign pressure, particularly if this pressure came from cultures considered to be traditional, such as the Latin American.

Second, scientific development occurs as the result of an external thrust that generally stems from a "more advanced" cultural influence. U.S. philanthropic agencies saw scientific development in the United States as a result of the German influence brought to the country by nineteenth-century scholars who returned to their country after spending some time in Germany. Similarly, the mid-nineteenth-century North American influence was thought to have encouraged Latin American scientific structures.

Third, scientific development was seen as the result of elite actions. The main emphasis of philanthropic agencies was either to limit the number of researchers or to focus agency efforts on a selected number of research centers and medical schools. It was reasoned that the multiplier effects of scientific activity would emerge spontaneously. The "concentration-of-elites" model did not consider a mechanism through which the promoted scientific-organization models could be extended; therefore, philanthropic agencies overlooked the problem of adaptation.

This idea was strongly influenced by North America's historical experience, in which competition and individualism were deeply rooted in academic life and in which prescribing standards of quality was considered to be enough to encourage emulation.

These assumptions resulted from an elitist interpretation of the scientific-development experience in Western Europe and the United States. This interpretation barely noticed the particular factors and dynamics that made it possible for scientific excellence to take place in the periphery. The history of the study of twentieth-century Latin American physiology and microbiology is particularly illustrative in this regard and includes a number of important factors.

Concentration of Institutional and Human Resources

One of the main problems that the Latin American scientific communities have to face concerns the size of the group. A small number of researchers is normally not enough for even the minimal specialization and administration that science implies. The most remarkable examples of Latin American scientific research have solved this problem by concentrating on just a few centers rather than promoting competitiveness. This approach has brought the small number of researchers together and has also produced a thematic concentration. As a consequence, the basic sciences have not developed uniformly or at the same rate; rather, basic research has become progressively more sophisticated as second- or third-generation scientists at the research centers begin to diversify the original scientific program.

Two cases illustrate this resource concentration: the study of Peruvian and Argentine physiology. The first experiments in Peruvian physiology were performed at the Instituto de Biología Andina (Andean Biology Institute), created in the early 1930s and affiliated with the Universidad Nacional Mayor de San Marcos medical school. Such studies helped describe the Andean native's characteristics and verify the physical and physiological mechanisms through which human beings adapt to altitude. The research program was later extended to include the biochemical, genetic, hematological, and cardiological aspects of altitude adaptation.[3] Expanded research was conducted by students of the first generation of researchers, who extended what they had been taught to include other disciplines. In this way, the study of physiology at altitude was useful not only for the emergence of experimental science in Peru but also for the promotion of other disciplines.

In 1919, physiologist Bernardo Houssay was appointed to the physiology chair at the Universidad de Buenos Aires, where, in a short time, he was able to create the Instituto de Fisiología (Institute of Physiology).[4] Under Houssay's direction, the institute incorporated physiology, biochemistry, and biophysics into its program and also offered medicine, odontology, pharmacy, and science courses. By the early 1930s, the institute's library subscribed to over seventy international journals, some of which were the most important in the fields of physiology, biochemistry, biophysics, and pharmacology.[5] The most significant research work was related to diabetes mellitus and the functioning of the pituitary gland, works for which Houssay would receive a series of international distinctions, including the 1947 Nobel Prize in Physiology or Medicine.

The Combination of Basic and Applied Research

Successful Latin American research efforts have also included the repeated and constant claim of researchers that science is useful not only as a vehicle for cultural expression but also as an aid in responding to social needs. Justifying research that seems esoteric and unrelated to the needs of a developing country has always been difficult in Latin America. This accounts for the difference between scientific development in Latin America and that of other regions. The latter have seen the development of biological and exact sciences that either preceded the development of applied science or took place apart from it; in Latin America, the scientific areas that attained excellence managed to create, at an early stage, a special combination of basic and applied work. Biomedical sciences are logically equipped to do both. Even today, most Latin American scientific publications are on biomedical science, according to the most specialized indicators of scientific publications.[6]

The clinical applications of Latin American physiological studies and the potential application of microbiology to epidemiology and public health have helped justify such research in a scarcely developed scientific environment. It is important to note that these possibilities have been exaggerated on several occasions by the scientific community itself, and that the safety of applying clinically acquired knowledge has not always been assured. Nevertheless, the emphasis on practical applications has helped outline a strategy for scientific investigation.

Brazilian microbiology is an important case. Between 1902 and 1920, Oswaldo Cruz, from Rio de Janeiro, managed to transform a small local laboratory dedicated to the production of serum and vaccines into a microbiology institute where first-class research was carried out.[7] This transformation occurred at a moment when Brazilian cities were undergoing an intense reorganization of public-health structures. This reorganization was the result of three major factors: the germ theory of disease; the adoption of the French model of institutionalization as traced by the Pasteur Institute in Paris; and the serious increase of bubonic plague and yellow fever in the country.

Oswaldo Cruz and members of the institute led the campaign against these diseases and to modify the structures of public health in Rio de Janeiro and São Paulo. One of the institute's most remarkable contributions was the identification and study of a new clinical entity that attacked the rural zones: Chagas disease.[8]

The Nationalism That Drove Science

According to some authors, science should be free from any localist or nationalist influence.[9] This argument has helped condemn the most extreme cases, in which dictatorships have manipulated science. It has also helped promote scientific autonomy and decision-making power and create an ideology of the so-called purity of science.

The situation in the developing countries, however, has been different. The Latin American governments frequently have been deaf to researchers' pleas for greater support, and promotion of "uncontaminated" science has contributed to an esoteric and alienated image of science.

The cases in which science has been fully integrated into developing countries suggest that it created a realm of its own within the local culture by relating research activities to particular issues concerning the country's interests. Nationalism led scientists to demand that their countries have their own institutes, laboratories, and libraries, like those in the developed world. Nationalism manifested itself, although not exclusively, not only through the thematic selection of research topics—such as the study of native diseases—but also in the content of science itself. Such was the case in Peru, where the development of research dealing with altitude emerged as a reaction against the Anglo-Saxon expeditionaries' conclusions that the Andean native was "limited" because of the scarcity of oxygen in his environment. Peruvian researchers demonstrated the existence of physical and physiological mechanisms by which Andean men and women have adapted to altitude for centuries.[10]

The reasons for stressing Andeans' extraordinary physiological capabilities are better understood within a framework in which the recovery of Andean culture—known as indigenism—was an intellectual position, as during the 1920s and the early 1930s in Peru. Indigenism in Peru included areas of human activity such as social science, archaeology, and painting.

The integration of science and local culture occurred as part of an intensive campaign during which inhabitants of Peru's major cities participated in the discussion about the importance of the discipline that was the subject of the campaign. Discussions were printed in newspapers and nonspecialist magazines and were part of other forms of public debate. They were important in allowing the population to internalize the need for and importance of science. This internalization widened and legitimized research possibilities and eventually broadened support from specialized institutions.

The informational campaigns differed from the traditional models of technical assistance in which emphasis was placed on consolidating a particular community without much contact with other sectors. Moreover, the traditional model of assistance considered any discussion with the scientific sector to be a waste of time that could be used for research. The traditionalists assumed that the rationality and superiority of Western science would be perceived by the public and that Latin American society lacked any concept of disease within its immediate surroundings.

The Search for Clients

Based on her study of Oswaldo Cruz's experience in Brazil, Nancy Stepan proposed in the early 1980s that scientific institutions in the developed countries needed to gather a minimum number of clients in the national sectors requiring their services.[11] Such sectors normally include the government, which can afford research specifically related to development projects or to the expansion of the state's bureaucratic structure. In Brazil, the Instituto Oswaldo Cruz was hired to prepare health reports and carry out vaccination campaigns and other prevention activities that were related to the state's public works. In Peru, the Instituto de Biología Andina established regular contacts with three groups of clients:[12] the mining companies, which were particularly concerned with the miners' diseases and the effects of altitude on mining productivity; cattle breeders from the central mountain range, who used the institute's aid to increase efficiency in raising wool-bearing cattle, which were affected by altitude;

and finally, the airlines, such as Pan Am, which were interested in how pilots were affected by altitude.

The Use of Unsophisticated Research Technology

Philanthropic aid to Latin America's laboratories emphasized modern and sophisticated equipment. For instance, at one time, the laboratory specializing in high-altitude physiology at the Instituto de Biología Andina in Lima had the same equipment as Harvard University. This philanthropic philosophy had some merit, as it did not consider the peripheral laboratories as second-class centers that did not deserve the most modern equipment. In the long term, though, the utility of this kind of aid is doubtful for several reasons.

Among other problems that the agencies did not take into account were equipment maintenance, the consequences of not updating the equipment, the lack of depth in the theoretical discussion, and, finally, the frequent stagnation of scientific work. A false association emerged between modern research and the possession of costly and sophisticated equipment.

Latin America's successful scientific experiences have occurred in areas in which no sophisticated technology was needed. The main instruments for bacteriological studies, for instance, were the microscope and a minimum of specialized literature that was easily available in most Latin American cities in the early twentieth century. In the case of Peruvian altitude-related physiology, the Andes were a natural laboratory with the same advantages as the sophisticated pressure chambers built in the United States intended to replicate the effects of altitude.

Early Integration in the International Context

All successful scientific episodes on the periphery have occurred when the object of research was relatively new or when international competition was not intense. This process differs from the linear model suggested by Basalla, which considers the colonial scientific phase to be a period when scientists from developing countries did not make significant contributions to international science.

At the beginning of the twentieth century, Latin American bacteriologists made numerous significant contributions, mainly because their research was focused on certain native diseases unknown in Europe, such as Carrion's disease in Peru and Chagas disease in Brazil. There was little international competition in these areas, which allowed greater opportunities for researchers in the periphery to make internationally significant contributions. The Latin American scientists who pioneered in these areas identified new fields within the international scientific context that had greater relevance for their own countries as well.

Conclusions

What all of this suggests is that Latin American scientists have been extraordinarily adaptive to the hard conditions they have had to face in order to develop their careers. This is most clearly exhibited by the continent's scientific pioneers, who began to write the stories of institutional success. They were trained as researchers in first-rate institutions in Europe or the United States; nevertheless, when they returned home, they found a need to build a minimal infrastructure that would allow research to be continuously improved. Many, therefore, became scientific managers and businessmen, for example, Carlos Monge Medrano at the Instituto de Biología Andina in Peru and the remarkable physiologist Alejandro Rosenblueth in the physiology laboratories at the Instituto Nacional de Cardiología (National Institute of Cardiology) in Mexico.

In contrast, North American philanthropic programs from the 1930s through the 1950s rarely made adapting their models a priority. The Instituto de Biología Andina and the Instituto Nacional de Cardiología were only partially supported. These philanthropic programs concentrated on just a few centers and often rejected relations with the local culture and nationalist motivations and did not search for local clients. This lack of flexibility was symbolized by a certain rigidity related to keeping science pure and protected from external influences.

Research that considers the study of scientific development as a discipline in the history of science in Latin America and other nonindustrialized countries has emerged only recently. Researchers must propose alternative elements and patterns to the traditional opinion of the center regarding the scientific development of the great majority of the world's regions. The scientists I have discussed here constantly struggled against the two true misfortunes of Latin American scientific history: the lack of institutional continuity; and the stagnation of the original research programs. Thus, the history of successful science in Latin America is incomplete, a promise that is yet to be fulfilled. Reflection on the elements that are present in all cases of scientific excellence might help Latin America meet that promise.

Notes

1. Basalla, "The Spread."
2. There is considerable literature on the topic. A classic work is Sábato (ed.), *El pensamiento*.
3. Cueto, "Andean Biology."
4. "Houssay, Bernardo Alberto."
5. Houssay, "Institute of Physiology," pp. 1–11.
6. Garfield, "Mapping Science."
7. See Benchimol, *Manguinho*.
8. The analysis of Cruz's work appears in Stepan, *Beginnings*.
9. This point of view goes back to Polanyi's "The Republic of Science."
10. See Cueto, "Nacionalismo."
11. Stepan, *Beginnings*.
12. Cueto, "Andean Biology."

Bibliography

Basalla, G. "The Spread of Western Science." *Science* 156 (May 5, 1967): 611–622.
Benchimol, J. L. (coord.). *Manguinhos do sonho á vida: A ciência na Belle Époque*. Rio de Janeiro: Casa de Oswaldo Cruz, 1990.
Cueto, M. "Andean Biology in Peru: Scientific Styles in the Periphery." *Isis* 80, no. 304 (December 1989): 640–658.
———. "Nacionalismo y ciencias médicas: Los inicios de la investigación biomédica: 1900–1950." *Quipu, Revista Latinoamericana de Historia de la Ciencia y la Tecnología* 4, no. 3 (September–December 1987): 327–356.
Garfield, E. "Mapping Science in the Third World." *Science and Public Policy* 10 (1983): 112–127.
Houssay, B. H. "Institute of Physiology Faculty of Medical Sciences University of Buenos Aires." In *Methods and Problems of Medical Education*. New York: Rockefeller Foundation, 1932.
"Houssay, Bernardo Alberto." In *Dictionary of Scientific Biography*, Supp. I, C. C. Gillispie (ed.), vol. 15, p. 228. New York: Charles Scribner's Sons, 1978.
Polanyi, M. "The Republic of Science: Its Political and Economic Theory." In *Criteria for Scientific Development*, E. Shils (ed.). Cambridge, Mass.: MIT Press, 1968.
Sábato, J. (ed.). *El pensamiento en la problemática ciencia-tecnología-desarrollo-dependencia*. Buenos Aires: Paidós, 1975
Stepan, N. *The Beginnings of Brazilian Science: Oswaldo Cruz, Medical Research and Policy, 1890–1920*. New York: Columbia University Press, 1976.

CHAPTER 9

International Politics and the Development of the Exact Sciences in Latin America

REGIS CABRAL

Introduction

The purpose of this chapter is to motivate the reader to participate in the process of constructing the history of the exact sciences in Latin America and internationally. Given that we live in a time in which the world and basic concepts are being reorganized, I am taking this liberty. My subject requires us to define five elements. Four of them are explicit—development, the exact sciences, Latin America, and international politics—and one is implicit—the periodization of the magnificent and, to many, confusing history of the twentieth century. If these five terms are understood well, confusion should diminish; therefore I will first discuss the four explicit elements, then will proceed to the periodization issue. I will then present my version of the exact sciences and their interaction with international politics and how that interface affects Latin America's place on planet Earth. Rather than conclusions, I shall simply comment, for there remains much to do to rediscover our history; thus, any conclusions would be premature.[1]

Definitions

Development

The concept of development has been under attack lately. Its use has led to all kinds of contradictions, particularly in discussions of economics. To many, development means industrialization, that is, overcoming poverty by acquiring industrial production. Hence, what is developing is not yet developed. But a "development area" in England is a poor area with high unemployment—economically depressed.

Behind the concept of development lies the belief that history passes through certain stages, or moments. One is led to believe that the issue or society under study has, at a later moment, something more than it had in an earlier moment or that it is larger or that it has expanded more. To state that something is developing suggests an almost always utopian vision of what one wants the society to become. This means that, when someone talks about the development of the exact sciences in Latin America, there is an underlying image of what these sciences should be. To some, this development is like that in the United States, Europe, or Japan; to others, it is a search for knowledge tied to Latin America's indigenous roots. There are those who try to calculate how many scientists Latin America has, how much they publish and where. There are many other examples, but I must mention those who dream of a day when the world will consider the knowledge that is produced in Latin America as a great source of universal knowledge.

The Exact Sciences

Are the exact sciences developing in Latin America? There is some consensus that an exact science is one that has mathematical and quantitative foundations. In theory, descriptive or classifying knowledge does not constitute an exact science. The great problem in knowing what an exact science is, and what it is not, comes from the evolution of the sciences. When we state that one science is exact and that another one is not, we assume—whether we want to or not—that a certain hierarchy exists, that is, a scale on which some sciences occupy the top position and others are below it. For many, the more mathematical a science, the more exact it is. This happens even with the social sciences, which try to quantify themselves in order to achieve a higher status in the public eye.

This idea even affects how the development of Latin American science is perceived. If we view the history of science through the lens of mathematics or, more precisely, through quantification, then we want the region to have a great number of scientists, we want a great many Nobel Prizes, and we mainly want the international scientific community to mention—that is, to use—the scientific works of Latin American scientists. But such quantification tells us very little about the quality of this science; it tells us very little about the social and economic effects of this science in Latin America; and it tells us nothing about its exactness.

What, then, is an exact science? Is it something that is closer to technology? Certainly not, for few consider biology to be an exact science, and nowadays the greatest technological progress is occurring in related areas, such as genetic engineering. The solution to this part of the puzzle is to use common sense. Such a method is suggested frequently in works on the sociology of technology and science, which have proved that separating these two social activities is almost impossible. The same suggestion has been made regarding the so-called exact sciences: use common sense with flexibility. Thus, physics, chemistry, and astronomy can be considered as exact sciences, whether they are carried out in departments of engineering, governmental agencies, or private companies.

But I insist that such areas of knowledge should not be considered better or worse than others, or, because of their content or type of organization—most important—as necessary models for the development of other areas of knowledge. In order to illustrate the issues, I shall consider primarily the development of physics. This does not mean that other areas will be excluded entirely.

Latin America and International Politics

It is difficult to define the concept of Latin America. Two questions illustrate this difficulty: Is Florida a part of Latin America? What about Surinam? Neither the language nor the culture nor the existence of Latin American states can be considered as the unifying element. From the point of view of science and technology, this is a critical situation, for one of the most advanced specialized technological centers in the world is located on our continent: the French space center in Guyana. Most Mexicans, Brazilians, and Chileans, for instance, have trouble considering such scientific technological activity as Latin American, but in fact it is.

The concept of international politics connotes the existence of nations. In our case, we will focus on the nation-state that attempts to produce science, whether as a reaction or out of necessity. We must differentiate between discourse and practice. Almost all Latin American states include (or will include in the near future) the category of science as a part of their official discourse. But it is clear that only a few have generated scientific institutions or activities that withstand the economic and political pressures that are inherent to history. Also, a great deal of existing scientific production depends on international organizations.

This leads us to a classification of the main characters playing this scientific game of international relations. There exist scientists as individuals (who can perform other types of activities, including political); there also exist the national institutions, which can be independent or dependent on the state and which can be dedicated to research or teaching, or which can belong to professional associations. There exist the international institutions, which can be privately financed organizations or regional organizations, agencies with a global scope or multinational for-profit corporations. Finally, there are the states themselves and their agencies.

It is precisely from these players' actions or appearance on stage that it is possible to start periodizing the history of the relationship between the exact sciences and international politics in Latin America. The following is a description of three main periods. This does not mean, however, that other possible periodizations are excluded. I am concentrating on those periods in which the activity of Latin Americans working in the exact sciences participates in and contributes to the region's international relations.

First Period: The Transition from the Nineteenth to the Twentieth Century

The end of the nineteenth century and the beginning of the twentieth was marked by a major transition in the physical sciences. Newton's classical physics had reached its limits, and alternatives were already beginning to take shape. Planck's quantum theory and Einstein's theory of relativity emerged. The world in which the atom was indivisible was totally shattered when X-rays and radium and its properties were discovered.

During this transition period, there were important scientific-technological interactions—discussed so well by Lewis Pyenson. The United States became a gigantic presence and, even though this was already evident in Mexico and Central America, it began displacing the European powers, Spain and England, in the military and the economic fields, respectively. Europe, increasing its military power and preparing for a war no one thought would last long, never ceased using its scientific and technological influence to strengthen its presence in Latin America. Both France and Germany made their own efforts as the bureaucracy in each country sensed the North American menace. For many countries, such as Argentina, the roots of the exact sciences were introduced by "missionaries and functionaries" before 1914, according to Pyenson, who were sent by France and Germany.

Within this context in which science bore a distinctive mark of cultural superiority and as a part of its imperialist program, Germany sent its scientists to the Universidad de La Plata. First, Emil Bose arrived in Argentina, followed shortly thereafter by Richard Gans. La Plata had, until the 1920s, a level of theoretical physics that matched that of many European centers and that could be well said to be unsurpassed in Latin America. In spite of its waning with time, the nucleus of the great development in physics was present. It is enough to say that the incomparable Enrique Gaviola was a student of Gans's.[2]

The United States' efforts, which included sending astronomers to Argentina, during this period took the form of, primarily, the participation and reluctant support of pan-American science. The U.S. secretary of state considered pan-American science to be an instrument for consolidating Washington's position. Not all politicians shared this perception, hence the reluctance. Notwithstanding the fact that the U.S. delegation to the fourth Pan-American Scientific Congress—held from December 1908 to January 1909—arranged for the next congress to take place in Washington, the U.S. Congress rejected the funding authorization, thus delaying the next meeting for three years, until 1915.[3]

The typical position of Latin American society, largely influenced by positivism, was to adopt European science mechanically, as an instrument required for improving living conditions on the periphery. The main agents were foreign scientists like the Germans who made La Plata the center it was. They were institutionally linked to their home countries, so science's image in Latin America as far as international politics was concerned, was similar to that of a pawn in a gigantic and complex chess game where kings, queens, and bishops battled in the greatest confrontation in the history of humanity.

Second Period: The World Wars, 1914–1945

World war was considered until recently to be composed of three phases, which some authors considered discontinuous: World War I, the Depression, and World War II. In spite of its Eurocentrism, this position excluded many European conflicts, such as the one in Spain, and as made clear by the names given to the old periods, the approach was marked by European and U.S. developments. But if we were to consider the Pacific only, we would require another periodization, such as the Great Fifteen-year War of 1931–1945. This would also be inappropriate, however, for it would lead us to argue for the existence of a Great German War, which began in 1914 and ended in 1945, but that would only partially explain the situation.

Therefore, I shall stick to the new classification, indicating it as a period not only of great scientific development but also of great applications of science. As soon as difficulties arose in commercial interactions, forcing Latin America to seek industrialization, scientific applications during the great confrontation provided strong and sufficient arguments for those interested in development and who wanted to motivate their societies and those with power. The Great War began with a war of chemists and the weapons they developed, and it ended with a physicists' war and the tragedies in Hiroshima and Nagasaki. In Latin America, the Great War began with physics as only a cultural activity, with national and imported heroes who attempted to organize science by trying to undermine positivism's pernicious foundations. It ended with a movement to integrate physics into the region's social and economic development.

Technological dependence had during this period a highly negative effect on the development of physics throughout almost all Latin America. Therefore, the region is still primarily a passive receiver of European and U.S. developments. A typical case is illustrated by Venezuela under the Juan Vicente Gómez dictatorship (1908–1935). The Academia de Ciencias, Físicas, Matemáticas y Naturales (Physics, Mathematics, and Natural Sciences Academy), created in 1917, had very little to do with these disciplines. The foreign missions outside of the military sector—influenced by the French—and the sanitation sector—with the Rockefeller Foundation's great contribution—left no gains. The free oil exporting country, by asking little in exchange, saw very few gains in terms of scientific development.[4] The same happened in an area of great economic importance for Venezuela, the oil sector; a mere three engineers were sent abroad in 1930 to study.[5]

The Academia Brasileira de Ciências (Brazilian Academy of Science), on the other hand, founded on May 3, 1916, in Rio de Janeiro as the Sociedade Brasileira de Sciências (its current name dates from 1922), is an active group that directly confronts the positivist stagnation of Brazilian culture. Directly linked to the Associação Brasileira de Educação (Brazilian Education Association) and to the Escola Politécnica do Rio de Janeiro (Polytechnic School of Rio de Janeiro), and, at first, to French interests through the Instituto Franco Brasileiro de Estudios Avançados (Franco-Brazilian Institute of Advanced Studies), the academy became the place where positivism was defeated. This happened in 1925, during Einstein's visit to Brazil on his way to Argentina. In Sobral during a total solar eclipse, Einstein confirmed his theory on May 29, 1919. While traveling through South America, he visited Henrique Morize, chief of the Brazilian mission, who, with Andrew Crommelin, an Irishman, had put the measurements into practice.[6]

After Einstein's departure, the positivist members of the academy ran to the press to attack relativity. Amoroso Acosta, Teodoro Ramos, Roberto Marinho de Azevedo, and Álvaro Alberto da Mota e Silva, all with a sound mathematical base and knowledge of the theory of relativity, counterattacked. Following this success, this group, along with others who participated in the academy's activities, began in Rio de Janeiro to change and found institutions. This process was catalyzed by the appearance of the atomic bomb during the next period.

Meanwhile, in São Paulo, the combination of a national and an international dynamic brought to Brazil one of its greatest names: Gleb Wataghin. He gathered an internationally renowned team, which included Mario Schenberg, among others.[7] Wataghin was born in Odessa in 1899, and fled to Italy in 1919; there he studied physics and mathematics and contributed to the understanding of cosmic rays. In 1934, he became part of the Faculdade da Filosofia (Philosophy Department) at the Universidade de São Paulo.

The State of São Paulo wanted to build a university as a response to the military defeat suffered by the constitutionalist revolt in 1932. The state's elite wanted to recover its position through knowledge, culture, and science. Their instrument was the university, founded on January 25, 1934. They tried to bring in the very best of Europe—Heinrich Rheinholdt, Ettore Onorato, Fernand Braudel, and Claude Lévi-Strauss—but in the case of the exact sciences, the process took a European turn. The Italian government wanted to promote Mussolini's version of culture and so handed the task to mathematician Luigi Fantappié.[8]

But Wataghin was not a Fascist and, unlike Fantappié, he remained in Brazil when the new phase of the war began.[9] In addition, his pupils engaged in military research and designed, for instance, a Brazilian sonar model. One of the highlights of the period was Arthur H. Compton's visit, along with his "Cosmic Ray Expedition."[10] Compton was not the only one impressed by Wataghin's work group; it also had an impact on Harry Miller Jr. from the Rockefeller Foundation and also part of the expedition. The good work carried out there would bring benefits in the following period, when the foundation decided to support nuclear physics in São Paulo.

But in general, this time period was characterized by a constant lack of financing. One of Wataghin's greatest skills was occasionally to obtain funds from authorities who regarded academics with suspicion.[11] In Rio de Janeiro,

Bernhard Gross, from Stuttgart and with a PhD in cosmic rays, had been hired by the Instituto Nacional de Tecnologia (National Institute of Technology) but was suffering from poor working conditions. Under these terrible circumstances, a member of Gross's group, Joaquim Costa Ribeiro, of the Faculdade da Filosofia of the Universidade de Rio de Janeiro, discovered the thermodielectric effect in 1944.[12]

At that time the authorities—even those in charge of promoting physics—had only a narrow understanding of what making physics a living part of society could mean. In his history of physics in Argentina, Ramón Loyarte—an icon of physics in that country who dismissed Gans in the 1930s—wrote that he could not conceive of a development that was independent from Europe; Argentine physics should orbit the European center.[13]

Meanwhile, signs emerged that this was not to be a permanent situation. In 1936, for instance, Gaviola was already at work on shaping a new generation of Argentine physicists at the Observatorio Astronómico (Astronomical Observatory) of Córdoba. During this period, a giant of Latin American physics entered the scene in Mexico, Manuel Sandoval Vallarta.[14] Notice that I am writing about Mexico, since this is what happened. In 1917, he went to the Massachusetts Institute of Technology, where he confirmed through experiments Olivier Heaviside's operational calculus formulas. From 1925 to 1932, he worked on relativity and quantum mechanics, and next on cosmic rays. In 1943, he started working on improving research conditions in Mexico.[15] From 1943 to 1951, he was head of the Comisión Impulsora y Coordinadora de la Investigación Científica (Commission for the Support and Coordination of Scientific Research), which had little funding, and from 1944 to 1945, he directed the Instituto de Física (Physics Institute) of the Universidad Nacional Autónoma de México (National Autonomous University of Mexico). It was not until 1949 that he returned to his country permanently.

Because of his activities in the United States and Europe, Vallarta established a network of contacts that constituted, during the following period, the roots of Mexican physics' international relations. Amid all this and at the beginning of this period, Vallarta trained many Mexican students, including four important Mexican scientists: Alfredo Baños, who achieved the distinction of being named professor emeritus by the University of California at Los Angeles; Carlos Graef Fernández and Luis Enrique Erro, who founded and directed the Observatorio Astrofísico (Astrophysical Observatory) of Tonantzintla; and Marcos Moshinsky.

By the end of this period, it was evident that the internal dynamics of science's international relations had allowed at least three Latin American countries—Argentina, Brazil, and Mexico—to develop the initial conditions for developing physics programs. Were it not for international relations, Wataghin might well not have gone to Brazil and Gaviola might not have come under Gans's wing (he would not have been there). On Einstein's recommendation, Gaviola was granted a scholarship by the International Education Board to study at Johns Hopkins University. Vallarta also received international scholarships; in 1927, he was granted a Guggenheim to go to Berlin and study with Einstein and Schroedinger. The period closed with these physicists and many others committed to organizing the study of physics in their countries. The lack of funding limited what could be done. An external jolt was necessary to stir Latin American society into taking physics seriously. That jolt was the atomic bomb.

Third Period: The Cold War, 1945–1990

Few historical events have made science and physics so important a topic for discussion as Hiroshima and Nagasaki's destruction did. All of a sudden, subjects as esoteric as quantum mechanics and the effects of radiation became the topic of newspaper articles and society columns. Although the moral implications of this application of physics and its likely economic repercussions were discussed, another dimension took precedence: national security.[16] This was the result of the period during which Latin America was interfered with at the end of World War I.

The cold war resulted in multiple tragic interventions and interference with the development of physics on the continent. During this period, the United States considered anything that was not under its control to be a threat. I will not detail every case but will merely illustrate the situation. It is important to note that the United States did not always behave consistently and evenly, particularly as it became clear that the greatest competition did not come from the Soviet Union but from Germany and Japan.

Four of the scientists I mentioned above also illustrate reactions as the new era began. Sandoval Vallarta was willing to participate in Mexico's national life and quickly filled a series of bureaucratic and administrative positions. Also, he represented Mexico in a number of international organizations. In 1946, he became president of the United Nations' Atomic Energy Commission. His interest in atomic energy took many forms, even as a member of the Comisión Nacional de Energía Nuclear (National Nuclear Energy Commission) from 1956 to 1972.

Argentina decided to enter the atomic age beginning in the 1940s. It was subject to U.S. intervention, however, which caused a serious setback. At the beginning of 1947, Argentina, under Perón's direction, had difficulties with the United States because of its position during World War II. This crisis was intensified by the almost childish behavior of the U.S. ambassador in Argentina, Spruille Braden. Argentina's attempt to conquer the European wheat market did not help matters.

It was Gaviola who proposed the creation of a nuclear research institute. He was backed by Austrian physicist Guido Beck, former assistant to Nobel Prize winner Werner Heisenberg. The Argentine navy provided the financing, but Gaviola insisted that the institute remain under civilian control. To ensure civilian control, Gaviola invited Heisenberg to participate; he signed a contract accepting the offer. Meanwhile, around February 20, 1947, the British blocked Heisenberg's departure for Argentina, and the U.S. press started a violent campaign against Gaviola's plans.

U.S. efforts were reinforced by the country's open backing of the new president of Uruguay, anti-Peronist Thomas Beretta. This support was not only verbal; it also included strategic aircraft. Six strategic B-29 bombers of Group 97 from Smoky Hill Air Base, Arkansas, landed in Montevideo the day Beretta took office. These bombers were at the disposal of Buenos Aires just when Argentina was trying to start an independent nuclear program. We should remember that Bernard Baruch defended, before the United Nations, a U.S. project that included atomic reprisals against "illegal" atomic-energy programs.

For no apparent reason, the Argentine navy mothballed its nuclear program. Was it because of the nuclear threat?[17] Could Perón have been influenced by this incident when he decided to cut off Ronald Richter's tragicomic fusion program, so well described by Mario Mariscotti?[18]

This is not the only case of nuclear threat in Latin America. Besides the Cuban missile crisis in 1962, strategic forces actively backed the Central Intelligence Agency's operations in Guatemala in May 1954. According to the information I have, the incident was never investigated. We now know that the threat against Buenos Aires was only a bluff, because the B-29s were not equipped to carry bombs, and the number of bombs was low. In January 1947, the United States had only one atomic bomb. But this must also be understood within the context of the cold war. The United States wanted to maintain the illusion of in fact being a nuclear power until there was no need for the pretense.

It was not only Argentina that had to confront the United States. In 1951, the Conselho Nacional de Pesquisas (National Research Council,

CNPq) was created in Brazil. Its president and greatest promoter was Álvaro Alberto da Mota e Silva, who had by this time been promoted to admiral. The council's main objective was to develop a nuclear program. The financing for the program would come from selling strategic minerals, including monazite. In the beginning there existed the illusion that collaboration with the United States would be possible, but it soon became clear that this was not going to happen. Secretly, da Mota e Silva and the council, backed by Pres. Getúlio Vargas, began a cooperative program with Germany. This was illegal, for Germany was still occupied by the Allies.

Once they finished their projects, the German scientists asked the occupation government for permission to produce the equipment; it was denied.[19] A fantastic meeting took place among da Mota e Silva, the German scientist responsible for the project, Groth, and the Allied high commissioner (the equivalent of an occupation governor) in Germany, James Conant. To da Mota e Silva, the United States' treatment of an ally—Brazil—was unacceptable, but permission was still denied. Admiral da Mota e Silva asked Conant a basic question: "What privilege or monopoly did the Creator grant you to allow this?"[20] The best Conant could do was propose a meeting with the Atomic Energy Commission of the United States. But the answer was predictable, and Groth commented to da Mota e Silva, "We will do anything you want. Brazil will have its ultracentrifuges just as we will have ours."[21]

While the Germans worked in secret, da Mota e Silva signed an agreement with France, with Francis Perrin of the Commissariat à l'Energie Atomique (French Atomic Energy Commission), and with Mathiessant of the Société des Produits Chimiques des Terres Rares (Rare Elements Chemical Products Society), the company that built the factory for the famous yellow cake of uranium in Brazil.[22]

On January 21, 1954, the Banco do Brasil (Bank of Brazil) deposited eighty thousand dollars in the Banco Alemão da América do Sul (German Bank for South America) as payment for the centrifuges. Groth and Bayerie were in charge of the German part of the project, which involved fourteen factories working in secret.

The Germans completed the machines, but before they could ship them, they were confiscated on Conant's orders.[23] On July 25, da Mota e Silva went to Germany to try to recover the centrifuges, but with no success. The intervention was a lot wider than just atomic energy, and, as we know, Getúlio Vargas committed suicide on August 24. Afterwards, the CNPq was reorganized, forcing Da Mota e Silva to resign and placing Brazil's nuclear program under the control of Atoms for Peace, in the United States.

Lest one think the intervention was widely supported in the United States, the North American scientific community—Oppenheimer, for instance—supported the Brazilian project. It is also important to remember the tremendous support the Rockefeller Foundation gave to Wataghin's group for training personnel and for buying the necessary equipment in the United States. The Universidade de São Paulo group could not have continued without U.S. support.

This kind of support was also given to other countries. In 1958, the Facultad de Ciencias (Science Department) at the Universidad Central de Venezuela moved to the forefront of modern science. The physics community was rather small and had a theoretical focus. A lack of laboratories did not allow suitable training. The first professors in the department were European, Argentine, and Brazilian.[24]

The following year, the Instituto Venezolano de Investigación Científica (Venezuelan Scientific Research Institute) was created, followed by the Consejo Nacional de Investigaciones Científicas y Técnicas (National Council of Scientific and Technical Research, CONICIT) in 1967.[25] In 1963, a science school was created at the Universidad de Oriente, primarily because of its international relationship with the University of Kansas under the umbrella of a Ford Foundation program. This represents the definitive legitimization of the North American model in Venezuela, which was also established in the Facultad de Ciencias at the Universidad de los Andes in Mérida.

Note that international relationships did not exist exclusively between North and South, which were affected by local expressions of the cold war. For instance, a good number of Argentines migrated to Venezuela, including physicist Manuel Bemporad, who became the first director of the Escuela de Física y Matemática (School of Physics and Mathematics).

Comments: Where Are We Heading?

Perhaps it is time to ask whether it is possible to create a place for Latin America.[26] The answer to this question is related in a basic way to the question da Mota e Silva posed to Conant: "What privilege or monopoly did the Creator grant you to allow this?" Who grants himself or herself the right of possessing or having knowledge? The great tension among nations stems from unequal access to and production of knowledge.

But as we all know, it is impossible for a society to produce everything in every field of knowledge. The balance must be dynamic. It is not necessary for everyone to do the same thing and in the same way, but the playing field must be level. As soon as a nation refuses to work on a level parallel to that of other nations—according to Boris Yeltsin—conflicts arise. Within the context of scientific relations, it is the level playing field that Latin Americans have been denied. But it is Latin Americans themselves who deny themselves equal opportunity by insisting on a lopsided orbit around Europe or the United States. Therefore, recovering the history of these international relationships is fundamental. It is from this recovery of history that da Mota e Silva's contribution and personality, and those of many others, flow.

Once again, Latin America finds itself in a period of historical transition. Let us hope that the region does not succumb to a new Great War or a new cold war. But this will be possible only if Latin America controls its own past, its own history, the necessary tool for equal relations.

Notes

1. This chapter was written in Portuguese, but this version is based on the Spanish translation of the original.
2. Pyenson, "*In Partibus Infidelium*," pp. 253–303. A wider view of science's role in German imperialist politics can be found in idem, *Cultural Imperialism*.
3. Sagasti and Pávez, "Ciencia y tecnología."
4. Freites, "La ciencia."
5. Vessuri, "The Implantation," pp. 107–123.
6. Morize, *Observatório astronómico*.
7. Schwartzman, *A Space*. Interviews with the main Brazilian physicists of the time can be found at the Centro para Pesquisa e Documentação da História Contemporânea Brasileira (Center for Research and Documentation of Contemporary Brazilian History) of the Getúlio Vargas Foundation in Rio de Janeiro. There are abstracts of the interviews, with biographical information, at CPDOC, *História*, pp. 130–131.
8. See Olivera de Castro, "A matemática," vol. 1, pp. 70–71.
9. Goldemberg, "100 anos"; Leite Lopes, "La física nuclear," pp. 30–64; Sant'Anna, *Ciência*; Schwartzman, *Formação*, pp. 224–226, 251–264, and 453.
10. Academia Brasileira de Ciências, *Symposium*.
11. For how Wataghin got money from the State of São Paulo's auditor, Adhemar de Barros, see Schwartzman, *Formação*, p. 262.
12. According to Harry Miller Jr., of the Rockefeller Foundation; quoted in ibid., p. 180.
13. Ramón Loyarte, quoted in Saldaña, "Marcos conceptuales," p. 63.
14. Cabral, "Sandoval Vallarta."
15. For Sandoval Vallarta's biography, see Sandoval Vallarta, *Obra científica*; idem, "Reminiscencias"; Mondragón and Barnés, "Introducción," pp. xi–xvi; Cruz Manjarrez, *Reseña*; Lozano, García-Colín, Calles, and Ridaura, "Historia."
16. Cabral, "Mexican Reactions"; idem, "Cultural Dimension"; Mariscotti, "The Bizarre Origin," pp. 16–24.

17. Cabral, "Ameaças."
18. Mariscotti, *El secreto atómico*; Cabral, "The Perón-Richter Fusion Program," pp. 77–106.
19. Brazil, *Relatório*, p. 105.
20. Ibid., p. 103.
21. Ibid., p. 104.
22. Bandeira, *Presença*, p.104. This book is rich in information and is highly recommended.
23. Távora, *Uma vida*, p. 23. Conant's attitude is described in Bandeira, *Presença*, pp. 359–360. See also Archer, *Política*, p. 7; idem, *Segundo depoimento*, p. 11; Álvaro Alberto da Mota e Silva to Getúlio Vargas, Secret, July 25, 1954, File 1954, Archivo Getúlio Vargas.
24. Vessuri, "El proceso."
25. Vessuri, "The Implantation," pp. 107–123.
26. Saldaña, "Marcos conceptuales," pp. 57–80.

Bibliography

Archives

Arquivo Getúlio Vargas. Rio de Janeiro.

Secondary Sources

Academia Brasileira de Ciências. *Symposium sobre Raios Cósmicos*. Rio de Janeiro, August 4–8, 1941. Rio de Janeiro: Imprensa Nacional, 1943.
Archer, R. "Política nacional de energia atômica." Speech made to Câmara de Deputados, June 6, 1956. Rio de Janeiro: Imprensa Nacional, 1956.
———. "Segundo depoimento sobre o problema da energia nuclear no Brasil." Speech made to Câmara de Deputados, November 9, 1967. Brasilia: Imprensa Nacional, 1967.
Bandeira, M. *Presença dos Estados Unidos no Brasil (Dois séculos de história*. Rio de Janeiro: Civilização Brasileira, 1973.
Brazil. Comissão Parlamentar de Inquérito para Proceder as Investigações sobre o Problema de Energia Atômica (CPI). *Relatório. 1959*. In *As razões do nacionalismo*, D. Salles (ed.). São Paulo: Fulgor, 1959.
Cabral, R. "Ameaças norte americanas contra América Latina: O caso da Argentina, 1947." *Ciência e Cultura* 40 (1988): 656–658.
———. "Cultural Dimension of the Latin American Nuclear Debate." In *The Nuclear Debate in Latin America*, R. Cabral (ed.). Göteborg, Sweden: University of Göteborg, 1990.
———. "The Mexican Reactions to the Hiroshima and Nagasaki Tragedies of 1945." *Quipu, Revista Latinoamericana de Historia de las Ciencias y la Tecnología* 4, no. 1 (1987): 81–118.
———. "The Perón-Richter Fusion Program, 1948–1953." In *Cross Cultural Diffusion of Science*, J. J. Saldaña (ed.). Cuadernos de Quipu, Series 2. Mexico City, 1987.

———. "Sandoval Vallarta: As condições de validade da macromecánica, e estrutura conceitual da mecánica." *Quipu, Revista Latinoamericana de Historia de las Ciencias y la Tecnología* 5, no. 3 (1988): 327–337.

Centro de Pesquisa e Documentação de História Contemporâneo (CPDOC). *História da ciência no Brasil: Acervo de depoimentos.* Rio de Janeiro: Financiadora de Estudos y Projetos, 1984.

Cruz Manjarrez, H. *Reseña histórica del Instituto de Física: Primera etapa, 1938–1953.* Mexico City: Universidad Nacional Autónoma de México, 1975.

———. *Reseña histórica del Instituto de Física: Segunda etapa, 1953–1970.* Mexico City: Universidad Nacional Autónoma de México, 1976.

Freites, Y. "La ciencia en la época del geomecismo." *Quipu, Revista Latinoamericana de Historia de las Ciencias y la Tecnología* 4, no. 2 (1987): 213–215.

Goldemberg, J. "100 años de física." Unpublished. Instituto de Física, Universidade de São Paulo, São Paulo, 1974.

Leite Lopes, J. "La física nuclear en Brasil: Los primeros veinte años." In *La ciencia y el dilema de América Latina: Dependencia o liberación*, J. L. Lopes (ed.). Mexico City: Siglo XXI, 1972.

Lozano, J. M.; L. García-Colín; A. Calles; and R. Ridaura. "Historia de la Sociedad Mexicana de Física." *Revista Mexicana de Física* 28 (1982): 277–293.

Mariscotti, M. "The Bizarre Origin of Atomic Energy in Argentina." In *The Nuclear Debate in Latin America*, R. Cabral (ed.). Göteborg, Sweden: University of Göteborg, 1990.

———. *El secreto atómico de Huemul: Crónica del origen de la energía atómica en la Argentina.* Buenos Aires: Sudamericana-Planeta, 1985.

Mondragón, A., and D. Barnés. "Introducción." In *Obra científica*, by M. Sandoval Vallarta. Mexico City: Universidad Nacional Autónoma de México/Instituto Nacional de Energía Nuclear, 1978.

Morize, H. *Observatório astronómico: Um século de história (1827–1927).* Rio de Janeiro: Salamandra, 1987.

Olivera de Castro, F. M. de. "A matemática no Brasil." In *As ciências no Brasil*, vol. 1, F. Azevedo (ed.). Rio de Janeiro, 1956.

Pyenson, L. *Cultural Imperialism and Exact Sciences: German Expansion Overseas, 1900–1930.* New York: Peter Lang, 1985.

———. "*In Partibus Infidelium*: Imperialist Rivalries and Exact Sciences in Early Twentieth-century Argentina." *Quipu, Revista Latinoamericana de Historia de las Ciencias y la Tecnología* 1, no. 2 (1984): 253–303.

Sagasti F. R., and A. Pávez. "Ciencia y tecnología en América Latina a principios del siglo XX: Primer Congreso Científico Panamericano." *Quipu, Revista Latinoamericana de Historia de las Ciencias y la Tecnología* 6, no. 2 (1989): 189–216.

Saldaña, J. J. "Marcos conceptuales de la historia de las ciencias en Latinoamérica: Positivismo y economicismo." In *El perfil de la ciencia en América Latina*, J. J. Saldaña (ed.). Cuadernos de Quipu, Series 1, Mexico City, 1986.

Sandoval Vallarta, M. *Obra científica.* Mexico City: Universidad Nacional Autónoma de México/Instituto Nacional de Energía Nuclear, 1978.

———. "Reminiscencias." *Naturaleza*, no. 4 (1973): 170–175.

Sant'Anna, V. M. *Ciência e sociedade no Brasil.* São Paulo: Símbolo, 1978.

Schwartzman, S. *Formação da comunidade científica no Brasil.* Rio de Janeiro: Editora Nacional, 1979.

———. *A Space for Science: The Development of the Scientific Community in Brazil.* Dordrecht: D. Reidel, 1991.

Távora, J. *Uma vida e muitas lutas.* Rio de Janeiro: José Olympio, 1976.

Vessuri, H. M. C. "The Implantation and Development of Modern Science in Venezuela and Its Social Implications." In *Cross Cultural Diffusion of Science,* J. J. Saldaña (ed.). Cuadernos de Quipu, Series 2. Mexico City, 1987.

———. "El proceso de profesionalización de la ciencia venezolana: La Facultad de Ciencias de la Universidad Central de Venezuela." *Quipu, Revista Latinoamericana de Historia de las Ciencias y la Tecnología* 4, no. 2 (May–August 1987): 253–281.